# INTRODUCTORY SYSTEMS ANALYSIS FOR PROCESS ENGINEERS

## BUTTERWORTH–HEINEMANN SERIES IN CHEMICAL ENGINEERING



**SERIES TITLES**

Bubble Wake Dynamics in Liquids and Liquid-Solid Suspensions *Liang-Shih Fan and Katsumi Tsuchiya*

Chemical Process Equipment: Selection and Design *Stanley M. Walas*

Chemical Process Structures and Information Flows *Richard S. H. Mah*

Computational Methods for Process Simulation *W. Fred Ramirez*

Constitutive Equations for Polymer Melts and Solutions *Ronald G. Larson*

Fundamental Process Control *David M. Prett and Carlos E. Garcia*

Gas-Liquid-Solid Fluidization Engineering *Liang-Shih Fan*

Gas Separation by Adsorption Process *Ralph T. Yang*

Granular Filtration of Aerosols and Hydrosols *Chi Tien*

Heterogeneous Reactor Design *Hong H. Lee*

Introduction to Microhydrodynamics *Sangtae Kim and Seppo J. Karrila*

Introductory Systems Analysis for Process Engineers *E.B. Nauman*

Molecular Thermodynamics of Nonideal Fluids *Lloyd L. Lee*

Phase Equilibria in Chemical Engineering *Stanley M. Walas*

Physicochemical Hydrodynamics: An Introduction *Ronald F. Probstein*

Transport Processes in Chemically Reacting Flow Systems *Daniel E. Rosner*

Viscous Flow: The Practical Use of Theory *Stuart W. Churchill*

# INTRODUCTORY SYSTEMS ANALYSIS FOR PROCESS ENGINEERS

**E. B. NAUMAN**
The Isermann Department of
Chemical Engineering
Rensselaer Polytechnic Institute
Troy, NY

**Butterworth - Heinemann**
Boston London Singapore Sydney Toronto Wellington

Recognizing the importance of preserving what has been written, it is the policy of Butterworth–Heinemann to have the books it publishes printed on acid-free paper, and we exert our best efforts to that end.

**Library of Congress Cataloguing in Publication Data**

Nauman, E.B.
Introductory Systems Analysis for Process Engineers / E.B. Nauman.
p. cm.
Includes index.
ISBN 0-409-90254-3
1. Chemical process control. 2. System analysis. I. Title.
TP155.75.N38 1990
660'.2815--dc20 90-1989
CIP

**British Library Cataloguing in Publication Data**

Nauman, E.B.
Introductory Systems Analysis for Process Engineers. / E.B. Nauman
1. Linear systems. Analysis
I. Title
003.74

ISBN 0-409-90254-3

Butterworth–Heinemann
80 Montvale Avenue
Stoneham, MA 02180

10 9 8 7 6 5 4 3 2 1
Printed in the United States of America

# CONTENTS

# PREFACE

Linear systems analysis is a field originated by electrical engineers and still largely dominated by them. However, the underlying mathematics are common to most branches of engineering, and particularly to process disciplines such as chemical, environmental, and materials engineering. Process engineers need the mathematical methods but would prefer to learn them in a context more relevant than the analysis of electrical circuits.

This book provides such a context. The emphasis is on dynamic models derived using unsteady-state material and energy balances. Examples include chemical reactions, heat and mass transfer, and residence time distributions in flow systems. Also included are brief chapters on electrical circuits and dynamic mechanical systems. These continue the tradition that process engineering, and particularly chemical engineering, is the most broadly based technological discipline.

This text treats traditional topics in linear systems analysis, including Laplace transforms (Chapter 4), their application to sets of ODEs with constant coefficients (Chapter 5), and to network analysis (Chapter 9). Also included is a review of classical solution techniques (Chapter 3) and an introduction to matrix methods and rudimentary stability theory (Chapter 11). Less traditional is the emphasis on numerical solution techniques (Chapters 1, 3, and 12) and the nonlinear examples scattered throughout the text. Particularly suited to process engineers is the network analysis of flow systems subject to material balance constraints (Chapter 9), the subsequent use of these methods to determine residence time distributions (Chapter 10), and a brief introduction to the application of Laplace transforms to PDEs (Chapter 5).

This book is intended as an undergraduate text for junior or senior year process engineers. It provides the mathematics needed for more advanced courses in process control, chemical reaction engineering, and process design. The course is taught to first semester juniors at Rensselaer and, for the chemical engineers, is immediately followed by a course in process control. We manage to include all twelve chapters in a single, fairly busy semester. The simulation examples in Chapter 12 are typically inserted at various points throughout the semester. The other chapters are taught in sequence.

Relatively few schools have a course like Rensselaer's. More typical is to include the necessary mathematics as part of more advanced courses. For such schools, the book may be used as a supplemental text which collects all the bits and pieces in a single place. The book may also be useful to practicing engineers or beginning graduate students who desire a gentle review or even introduction to topics they will need throughout their careers.

Acknowledgements are due Rose Primett for typing the manuscript, my teaching assistants John Furno and Marci Reynolds who debugged it, Gus Aris and others who evaluated it, and Greg Franklin who accepted it for publication. Ordinarily, I would also thank those undergraduates who suffered through the early typos. However, they seemed grateful enough to avoid a semester that would otherwise have been filled with springs, dashpots, and RCL circuits. Process engineers think stirred tanks are more fun.

E. B. Nauman
Troy, New York
May, 1990

# INTRODUCTORY SYSTEMS ANALYSIS FOR PROCESS ENGINEERS

# CHAPTER 1

# SETS OF ORDINARY DIFFERENTIAL EQUATIONS

## 1.1 THE CANONICAL PROBLEM

This book analyzes a wide variety of engineering systems with examples chosen from chemical, environmental, electrical and mechanical engineering. The treatment has two purposes:

> To illustrate the dynamic behavior of systems that are important to the various engineering disciplines.
>
> To introduce the mathematical methods needed to analyze dynamic systems in general.

The second purpose is the more important. Few engineers will spend their entire careers in a single field. Most will have several changes in jobs and areas of application. Some will change their entire branch of engineering or will help define new, emerging branches. It seems more important to learn methods for analyzing new systems than facts about existing systems.

The analysis of engineering systems has three main components: modelling, solution, and verification. **Modelling** is the application of scientific and engineering principles to develop a set of equations believed to describe a physical system. The basic principles can come from any scientific or engineering discipline. Chemical engineers use the concepts of mass and energy balances, thermodynamics, transport phenomena, and reaction engineering. All of these topics will be covered along with topics from other branches of engineering. This book teaches modelling primarily through examples rather than a formal methodology. **Solution** is a mathematical procedure which produces either a formula (an analytical solution) or a set of numbers (a numerical solution). After solution comes **verification**. Does the real system behave as the model predicts? If not, modify the model until its predictions become more realistic.

Of the three steps in analysis, the solution step is the most easily generalized. The systems will be different and will be governed by different physical principles. The verification step will require experimental techniques appropriate to a specific field. However, once stated in mathematical form, many models look alike and can be solved in the same ways.

A generally perceived hierarchy of engineering models has the sequence

1. Single Algebraic Equations
2. Sets of Algebraic Equations
3. Single, Ordinary Differential Equations
4. Sets of Ordinary Differential Equations
5. Single, Partial Differential Equations
6. Sets of Partial Differential Equations
7. Monte Carlo and Stochastic Models.

The reader will be familiar with at least the first and second levels in this sequence through previous engineering and mathematics courses. Single algebraic equations arise from the steady state material balances that chemical engineering students study in courses called "Chemical Process Principles," "Material and Energy Balances," or "Stoichiometry." Sets of algebraic equations arise in multi-component, steady state systems such as the multi-component flash evaporations often studied in the same course. Analytical solutions are possible for simple situations such as single quadratic equations or sets of first order equations. Numerical techniques, such as Newton's method, are generally required for nonlinear equations, whether single or multiple.

The present book is primarily concerned with model levels three and four. All the solutions from these models can be found from the solution to a single, canonical problem! The word **canonical** means representative or standard. Almost every solution technique in this book is concerned with a single, canonical problem: Solve a set of simultaneous, first order ODEs subject to specified initial values. The set to be solved is

$$
\begin{aligned}
\frac{da}{dt} &= R_A\,(a,b,c,...) \\
\frac{db}{dt} &= R_B\,(a,b,c,...) \qquad\qquad (1.1) \\
\frac{dc}{dt} &= R_C\,(a,b,c,...) \\
&\vdots
\end{aligned}
$$

and the initial conditions are

$$\begin{aligned} a &= a_0 \text{ at } t = 0 \\ b &= b_0 \text{ at } t = 0 \\ c &= c_0 \text{ at } t = 0 \\ &\vdots \end{aligned} \tag{1.2}$$

If you can solve Equations 1.1 (subject to Equations 1.2 as initial conditions), you can solve a great many problems of engineering importance. Indeed, you can solve all the problems at model levels three and four.

Sets of first order ODEs are important for several reasons:

A. They frequently arise directly in the analysis of engineering systems. The set of Equations 1.1 may, for example, represent the concentrations of the various components in a chemical reactor.

B. Single higher order ODEs can always be expressed as an equivalent set of simultaneous, first order ODEs. Thus, model level three is always a subset of level four. (The converse, reducing a set of simultaneous ODEs to a single, higher order ODE, is not always possible.)

C. Some solution techniques for partial differential equations (model levels five and six) approximate the system as a set of simultaneous ODE's and thus use the techniques described in this book.

**EXAMPLE 1.1** Express the following third order ODE as a set of simultaneous, first order ODEs:

$$\frac{d^3x}{dt^3} + \frac{x^2}{t}\left(\frac{d^2x}{dt^2}\right) + x^3\left(\frac{dx}{dt}\right)^3 + \alpha e^{-\beta x} = \frac{x}{t^3}$$

where $\alpha$ and $\beta$ are constants.

**Solution:** The basic variable is x. We use its derivatives to define to ancillary variables.

$$y = \frac{dx}{dt}$$

$$z = \frac{dy}{dt} = \frac{d^2x}{dt^2}$$

so that

$$\frac{dz}{dt} = \frac{d^3x}{dt^3}$$

Substituting into the original equation

$$\frac{dz}{dt} + \frac{x^2z^2}{t} + x^3y^3 + \alpha e^{-\beta x} = \frac{x}{t^3}$$

The desired set of ODEs is

$$\frac{dz}{dt} = \frac{x}{t^3} - \frac{x^2z^2}{t} - x^3y^3 - \alpha e^{-\beta x}$$

$$\frac{dy}{dt} = z$$

$$\frac{dx}{dt} = y$$

---

The general approach is to introduce n-1 ancillary variables when the original equation is of order n. Thus, if an ODE contains $d^na/dt^n$ as its highest term, we define the ancillary variables $a_1, a_2, \ldots, a_{n-1}$ by

$$a_{n-1} = \frac{d^{n-1}a}{dt^{n-1}}$$

$$\vdots \qquad \vdots$$

$$a_2 = \frac{d^2a}{dt^2} \tag{1.3}$$

$$a_1 = \frac{da}{dt}$$

These can be written as

$$\frac{da}{dt} = a_1$$

$$\frac{da_1}{dt} = a_2$$

$$\frac{da_2}{dt} = a_3 \qquad (1.4)$$

$$\vdots$$

$$\frac{da_{n-2}}{dt} = a_{n-1}$$

The desired result is a set of n simultaneous, first order ODEs, with one ODE for each of the variables a, $a_1$, ..., $a_{n-1}$. Equations 1.4 provide n-1 of these in the variables a, $a_1$, ..., $a_{n-2}$, and we are lacking only the ODE for $a_{n-1}$. It can be obtained by substituting Equations 1.3 into the original, $n^{th}$ order ODE to eliminate all derivatives lower than $d^na/dt^n$. The original ODE will still contain the term for $d^na/dt^n$, but this is given by

$$\frac{d^na}{dt^n} = \frac{da_{n-1}}{dt} \qquad (1.5)$$

Thus the original ODE provides the first order ODE for $a_{n-1}$. The final result will be a set of n, first order ODEs, all but the last of which have the form of Equation 1.4.

The above procedure, with a slightly different notation, was used to solve Example 1.1. It can also be extended to sets of higher order ODEs.

**EXAMPLE 1.2** Express the following set of higher order ODEs as an equivalent set of first order ODEs:

$$D\frac{d^2a}{dx^2} + u\frac{da}{dx} - k_I ab = 0$$

$$D\frac{d^2b}{dx^2} + u\frac{db}{dx} - k_I ab - k_{II} b = 0$$

where D, u, $k_I$, and $k_{II}$ are constants.

**Solution:** Here n=2 so that we need only one ancillary variable for each original variable. These ancillary variables are defined using Equation 1.3

$$a_1 = \frac{da}{dx} \text{ and } b_1 = \frac{db}{dx} \tag{i}$$

Equation 1.5 gives

$$\frac{d^2a}{dx^2} = \frac{da_1}{dx} \text{ and } \frac{d^2b}{dx^2} = \frac{db_1}{dx}$$

Substituting into the original set gives

$$D\frac{da_1}{dx} + ua_1 - k_I ab = 0$$

and (ii)

$$D\frac{db_1}{dx} + ub_1 - k_I ab - k_{II} b = 0$$

The desired set of first order ODEs has four members given by Equations i and ii.

---

## 1.2 A NUMERICAL SOLUTION TECHNIQUE

The set of Equations 1.1, subject to the initial conditions of Equations 1.2, can always be solved by numerical methods. The simplest possible technique is known as **Euler's method** or **marching ahead**. This method relies on the definition of a derivative:

$$\frac{da}{dt} = \lim_{\Delta t \to 0} \frac{a(t + \Delta t) - a(t)}{\Delta t} \tag{1.6}$$

We will consider the case where $\Delta t$ is finite rather than infinitesimal. An approximation for a first derivative is

$$\frac{da}{dt} \approx \frac{a(t + \Delta t) - a(t)}{\Delta t} \tag{1.7}$$

Using this approximation for the lefthand side of Equations 1.1 gives

$$
\begin{aligned}
a(t+\Delta t) &= a(t) + \Delta t R_A \\
b(t+\Delta t) &= b(t) + \Delta t R_B \\
c(t+\Delta t) &= c(t) + \Delta t R_C \\
\vdots \qquad & \quad \vdots \qquad \vdots
\end{aligned}
\qquad (1.8)
$$

Here, $a(t+\Delta t)$ represents the value of a at the new time $t+\Delta t$, while $a(t)$ is the value at the old time t. The quantities $R_A$, $R_B$, $R_C$, ... are evaluated at the old time.

The initial conditions of Equations 1.2 give us the values of a,b,c,... at t = 0. The marching ahead algorithm is started at these initial values, which are then incremented using Equations 1.8. The step size of the independent variable, $\Delta t$, is a constant in simple forms of marching ahead but can be varied from step to step in more sophisticated forms.

Another way of writing Equations 1.8 is

$$
\begin{aligned}
a_{new} &= a_{old} + \Delta t\, R_A\,(a_{old}\ b_{old}, c_{old}, ..., t_{old},\ ) \\
b_{new} &= b_{old} + \Delta t\, R_B\,(a_{old}\ b_{old}, c_{old}, ..., t_{old},\ ) \\
c_{new} &= c_{old} + \Delta t\, R_C\,(a_{old}, b_{old}, c_{old}, ..., t_{old},\ ) \\
\vdots \qquad & \quad \vdots \qquad \vdots \\
t_{new} &= t_{old} + \Delta t
\end{aligned}
\qquad (1.9)
$$

**EXAMPLE 1.3** Determine a and b at time t = 1 from the following set of ODEs:

$$\frac{da}{dt} = -ab$$

$$\frac{db}{dt} = -ab - 0.1b$$

The initial conditions are that $a_o = b_o = 1$ at $t = 0$.

**Solution:** Pick $\Delta t = 1/4$ arbitrarily. Then apply Equations 1.9 to develop the following table.

| t | a | b | $R_A$ = -ab | $R_B$ = -ab-.1b |
|---|---|---|---|---|
| 0 | 1.000 | 1.000 | -1.000 | -1.100 |
| 0.25 | 0.750 | 0.725 | -0.544 | -0.616 |
| 0.50 | 0.614 | 0.571 | -0.351 | -0.408 |
| 0.75 | 0.526 | 0.469 | -0.247 | -0.294 |
| 1.00 | 0.465 | 0.396 | -0.184 | -0.223 |

Thus, approximate results are a = .465 and b = .396 at t = 1. Improved results can be obtained by decreasing $\Delta t$:

| $\Delta t$ = | 1/4 | 1/8 | 1/16 | 1/32 | 1/64 | 1/128 | 1/256 | 1/512 |
|---|---|---|---|---|---|---|---|---|
| a (t=1) | 0.465 | 0.492 | 0.504 | 0.510 | 0.513 | 0.514 | 0.515 | 0.515 |
| b (t=1) | 0.396 | 0.424 | 0.437 | 0.443 | 0.446 | 0.448 | 0.449 | 0.449 |

Hence, a = .515 and b = .449 are better approximations. In essence, we have taken the limit as $\Delta t$ approaches zero and thus have obtained a nearly exact solution to the original ODEs. The correct answers to high precision are a = .51562 and b = .44938.

The computing program needed to implement the marching ahead algorithm is extremely simple. The following BASIC language program is suitable for implementation on a PC:

```
        TFINAL = 1.0
        A0 = 1.0
        B0 = 1.0
1       INPUT N
        DT = TFINAL/N
        A = A0
        B = B0
        FOR  I = 1 TO N
                RA = -A*B
                RB = -A*B - 0.1*B
                A = A + DT*RA
                B = B + DT*RB
                T = T + DT
        NEXT I
        PRINT T, A, B,
        GO TO 1
```

## 1.3 DIMENSIONAL CONSISTENCY

Physically meaningful equations must have consistent units. The nth derivative of a with respect to x, $d^n a/dx^n$, has units of

$$\frac{[\text{a units}]}{[\text{x units}]^n}$$

Thus, if a is a concentration in moles/m$^3$ and x is a distance in m, $d^2a/dx^2$ has units of moles/m$^5$. Each additive term in an ODE must have the same units. Arguments of functions such as sin and exp must be dimensionless, as are the functions themselves.

**EXAMPLE 1.4** Is it possible for the ODE of Example 1.1 to be dimensionally consistent?

**Solution:**

$$\frac{d^3x}{dt^3} + \frac{x^2}{t}\left(\frac{d^2x}{dt^2}\right) + x^3\left(\frac{dx}{dt}\right)^3 + \alpha e^{-\beta x} = \frac{x}{t^3}$$

The corresponding units are:

$$\frac{\text{x units}}{[\text{t units}]^3} + \frac{[\text{x units}]^3}{[\text{t units}]^3} + \frac{[\text{x units}]^6}{[\text{t units}]^3} + [\alpha\ \text{units}] = \frac{\text{x units}}{[\text{t units}]^3}$$

The terms are inconsistent with respect to the x units and thus the overall equation is inconsistent except for the special case where x is a dimensionless variable. The constant $\beta$ must also be dimensionless since the argument of the exponential function must be dimensionless. The various terms are consistent with respect to the t units provided the constant $\alpha$ has units of [t units]$^{-3}$.

---

## 1.4 SCALING

It is common practice in engineering analysis, particularly in chemical engineering, to reduce physical variables to dimensionless forms. This is done by dividing the original variable by some characteristic value for that variable. Thus, a concentration is replaced by the dimensionless variable $a/a_0$, a length is replaced by x/L, and so on.

Putting equations in dimensionless form is useful as a means for generalizing numerical results. However, it should not be done too early in the analysis. Retaining dimensions during algebraic manipulations provides a good check on the results. Retaining dimensions also prevents the overly enthusiastic modeler from using functional forms that could not represent a real physical system. Thus, if x in Example 1.4 were a dimensioned variable, we would know immediately that the ODE could not represent a real system. The dimensions are inconsistent.

Once the dimensional consistency of the model equations has been confirmed, it is often useful to convert to dimensionless variables. This conversion is sometimes called scaling. It reduces the number of independent parameters in the model to the lowest possible number. It also allows numerical calculations to be generalized to include many separate cases.

**EXAMPLE 1.5** Scale the following differential equation:

$$\frac{da}{dt} = -k\,a$$

subject to $a = a_0$ at $t = 0$.

**Solution:** As written, this model contains one dependent variable (a), one independent variable (t), and two parameters ($a_0$, k). We first scale the dependent variable using its initial value. Define

$$\alpha = a/a_0$$

so that the model becomes

$$\frac{d\alpha}{dt} = -k\,\alpha$$

subject to $\alpha = 1$ at $t = 0$. Rewrite it as

$$\frac{d\alpha}{d(kt)} = -\alpha$$

and note that kt must be dimensionless since $\alpha$ is dimensionless. (We assume the original model to be dimensionally consistent.)

Thus, a dimensionless independent variable is

$$\tau = kt$$

and the model becomes

$$\frac{d\alpha}{d\tau} = -\alpha$$

subject to $\alpha = 1$ at $\tau = 0$. Converting this model to dimensionless form has eliminated all parameters. Thus a single numerical (or analytical) solution would apply for all possible values of $a_o$ and k. We would, of course, need to know $a_o$ and k to convert from the dimensionless variable $\alpha$ and $\tau$ back to the original variables a and t.

---

**EXAMPLE 1.6** Scale the following set of model equations and determine the minimum number of independent parameters.

$$\frac{da}{dt} = -k_I ab$$

$$\frac{db}{dt} = -k_I ab - k_{II} b \qquad \text{(i)}$$

subject to $a = a_o$ and $b = b_o$ at $t = 0$.

**Solution:** As written, the model contains one independent variable (t), two dependent variables (a,b), and four parameters ($k_I$, $k_{II}$, $a_o$, $b_o$). Scaling will replace a, b, and t with new, dimensionless variables but will not change the number of such variables. We first scale a and b using their initial values. The new dependent variables are:

$$\alpha = a/a_o, \qquad \beta = b/b_o$$

so that the model equations become

$$\frac{d\alpha}{dt} = -k_I b_o \alpha\beta$$

$$\frac{d\beta}{dt} = k_I a_o \alpha\beta - k_{II}\beta$$

subject to $\alpha = \beta = 1$ at $t = 0$. There are now only three parameters; $k_{II}$ and the products $k_I b_o$ and $k_I a_o$. However, it remains to convert t to dimensionless form. Rewrite the model equations as

$$\frac{1}{k_I b_o}\frac{d\alpha}{dt} = -\alpha\beta$$

$$\frac{1}{k_I b_o}\frac{d\beta}{dt} = -\left(\frac{a_o}{b_o}\right)\alpha\beta - \frac{k_{II}}{k_I b_o}\beta$$

Since $\alpha$ and $\beta$ are dimensionless, it follows that the product $k_I b_o t$ is dimensionless. Thus, a dimensionless independent variable is

$$\tau = k_I b_o t$$

The model becomes

$$\frac{d\alpha}{d\tau} = -\alpha\beta$$

$$\frac{d\beta}{d\tau} = -\left(\frac{a_o}{b_o}\right)\alpha\beta - \left(\frac{k_{II}}{k_I b_o}\right)\beta$$

subject to $\alpha = \beta = 1$ at $\tau = 0$. There are now only two parameters: $(a_o/b_o)$ and $(k_{II}/k_I b_o)$. These parameters are dimensionless numbers and they constitute the minimum set of independent parameters. Note that Example 1.3 solved this set for the case of $a_o/b_o = 1$ and $(k_{II}/k_I b_o) = 0.1$

---

The choice of scaling parameters is arbitrary; and in a complex set of equations, there may be several equally good ways of scaling. The exact form of the dimensionless parameters which remain after scaling will depend on the specific choice of scaling parameters. However, the final number of dimensionless parameters will be independent of the choice of the scaling parameters. Numerical results for the dimensionless variables will also depend on the choice of the scaling parameters, but will be identical when converted back to the original, dimensioned variables.

**EXAMPLE 1.7** Suppose a and b have the same dimensions. Repeat Example 1.6 using the following dimensionless variables:

$$\alpha = a/a_o, \quad \beta = b/a_o, \quad \tau = k_{II} t$$

**Solution:** Some algebra gives

$$\frac{d\alpha}{d\tau} = -\left(\frac{k_I a_o}{k_{II}}\right)\alpha\beta$$

$$\frac{d\beta}{d\tau} = -\left(\frac{k_I a_o}{k_{II}}\right)\alpha\beta - \beta$$

subject to $\alpha = 1$ and $\beta = b_o/a_o$ at $\tau = 0$.

These equations look somewhat different from those in Example 1.6. There are still two dimensionless parameters, but one of them now appears as an initial condition.

---

A distinction should be made between physically manipulatable parameters and the dimensionless parameters that arise from scaling model equations. The physical parameters, such as $a_o$, $b_o$, $k_I$, $k_{II}$ in the above examples, may indeed be independently varied by the experimenter. Often, such variation is desirable to confirm that the real system actually behaves as the model predicts. If it does, then the response of the system itself – when expressed in terms of dimensionless variables – will depend only on the dimensionless groups such as $(a_o/b_o)$ and $(k_I a_o/k_{II})$ that arise from the scaling.

## PROBLEMS

1.1 Express the following equations as an equivalent set of first order ODEs:

$$a^2\left(\frac{d^2a}{dx^2}\right)^2 + \left(\frac{db}{dx}\right)^4 + k_I a^2 b^2 = 0$$

$$b\left(\frac{d^2b}{dx^2}\right) + \left(\frac{da}{dx}\right)^2 + k_{II} a = 0$$

1.2 Suppose a and b are chemical concentrations and x is a length. Can these equations be dimensionally consistent?

1.3 Solve the equations of Example 1.6 for the case of $(a_o/b_o) = 0.8$, $(k_{II}/k_I b_o) = .2$. Find the value of $\tau$ for which $\alpha = 0.5$. Your result should be accurate to three significant figures.

1.4 Consider Equations i in Example 1.6 for the case of

$a_o = 4.2$ moles/m$^3$ $\quad$ $b_o = 2.8$ moles/m$^3$

$k_I = 2.6$ [consistent units] $\quad$ $k_{II} = 1.3\ s^{-1}$

a) What are the units of $k_I$?

b) Find t and $\tau = k_I b_o t$ for which $b/b_o = 0.1$.

c) Repeat b) using $a_o = 1.5$ moles/m$^3$, $b_o = 1.0$ moles/m$^3$, $k_{II} = 5.6$ s$^{-1}$, $k_I = 10$ [consistent units].

1.5 Reduce the following set of equations to a single ODE:

$$\frac{dx}{dt} = y$$

$$\frac{dy}{dt} = z$$

$$\frac{dz}{dt} + 3y^2 = 0$$

# CHAPTER 2

# SOME SYSTEM MODELS

## 2.1 TOTAL AND COMPONENT BALANCES

Chapter 1 covered some mathematical aspects of our overall subject. There, we introduced sets of first order ODEs, their dimensional analysis and scaling, and their numerical solution. We now consider how such sets of equations arise in engineering systems. There are a great many examples that could be treated. Mechanical engineers love their springs and dashpots, electrical engineers have a great fondness for RCL circuits, and chemical engineers rejoice in their stirred tank reactors. All of these example systems give rise to sets of ODEs. This chapter considers the stirred tank reactor illustrated below.

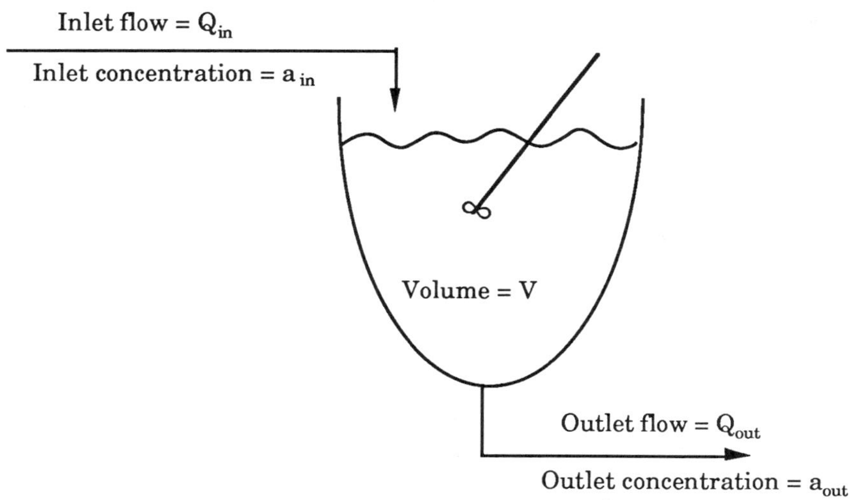

A general material balance is

$$\text{Input} + \text{Rate of Formation} = \text{Output} + \text{Rate of Accumulation} \quad (2.1)$$

For a non-reactive component in a fluid of constant density, the material balance gives

$$Q_{in} + 0 = Q_{out} + \frac{dV}{dt} \quad (2.2)$$

or

$$\frac{dV}{dt} = Q_{in} - Q_{out} \quad (2.3)$$

where all terms have units of [volume]/[time]. Equation 2.3 is a first order ODE for the system volume as a function of time. We suppose $Q_{in}$ and $Q_{out}$ are prescribed and that we also know an initial condition, say $V = V_0$ at $t = 0$. Then integration of Equation 2.3, either analytically or numerically, gives the system volume as a function of time, V(t).

A component balance for a reactive component is

$$Q_{in}a_{in} + VR_A = Q_{out}a_{out} + \frac{d(V\hat{a})}{dt} \quad (2.4)$$

where a is the concentration of component A and $R_A$ is its rate of formation per unit volume of reactor. The accumulation term, $d(V\hat{a})/dt$, depends on the average concentration of A within the reactor. We now suppose that the reactor is a **perfect mixer** so that the concentration at each point in the reactor is the same. This implies that $\hat{a} = a_{out}$ since the exit stream is taken from the perfectly mixed contents of the vessel. Further, the reaction rate function, $R_A$, is evaluated at the average concentrations, $R_A = R_A(a_{out}, b_{out}, \ldots)$. It is this assumption of perfect mixing that allows treatment of the stirred tank as a **lumped parameter system**. By this we mean that spatial variations within the stirred tank are ignored. Factors such as temperature, reaction rate, and component concentration may in fact vary from point to point within the system. Suppose for example that the reaction is exothermic but that there is cooling at the walls. A real reactor would be somewhat cooler near the walls and hotter near the center. It would also have higher reaction rates and higher concentrations of products near the center. Such real reactors are **distributed parameter systems** and are governed by partial differential equations. The assumption of **perfect mixing** allows a significant simplification. The concentration of a reactive component is assumed constant everywhere within the system and is governed by an ordinary differential equation:

$$Q_{in}a_{in} + VR_A(a_{out}, b_{out}, ...) = Q_{out}\ a_{out} + \frac{d(a_{out}\ V)}{dt} \qquad (2.5)$$

The initial condition is $a = a_0$ at $t = 0$. Equations 2.3 and 2.5, together with their initial conditions, constitute a complete description of a constant density, isothermal, unsteady state, variable volume, stirred tank reactor for a single component, $R_A = R_A(a_{out})$. For multiple components $R_A = R_A(a_{out}, b_{out}, ...)$, we add a version of Equation 2.5 for each component.

This book considers only four possible forms for $R_A$:

**Inert Tracer:**

$$R_A = 0 \qquad (2.6)$$

**First Order, Unimolecular:**

$$A = \rightarrow \text{Products},\ R_A = -ka \qquad (2.7)$$

Typical units for a are moles/m$^3$ and for $R_A$ are moles/m$^3$/hr. Then k will have units of hr$^{-1}$. All first order rate constants have units of reciprocal time.

**Second Order, One Reactant**

$$2A \rightarrow \text{Products},\ R_A = -2ka^2 \qquad (2.8)$$

The factor of 2 that appears in the rate equation is due to the stoichiometric coefficient on component A. See Section 2.3.2. Typical units on a and $R_A$ are moles/m$^3$ and moles/m$^3$/hr. Then the units on this k, will be hr$^{-1}$ m$^3$/mole. All second order rate constants have units of reciprocal time and reciprocal concentration.

**Second Order, Two Reactants**

$$A + B \rightarrow \text{Products},\ R_A = -kab \qquad (2.9)$$

The units on k are the same as in Equation 2.8, time$^{-1}$ concentration$^{-1}$. The student will learn about more general reactions and about the justification of and restrictions on the above rate equations in courses on chemical kinetics. For now, we take them as given.

## 2.2 STEADY STATE OPERATION

Steady state operation implies $Q_{in} - Q_{out} = dV/dt = 0$. At steady state, the reactor has volume V and volumetric flow rate Q. Steady state operation with respect to the chemical components implies $a_{in}$, $b_{in}$, ... are all constant.

$$Qa_{in} + VR_A (a_{out}, b_{out}, ...) = Qa_{out} + 0 \tag{2.10}$$

Note that the final term of Equation 2.4 vanishes at steady state. Thus the time derivative vanishes, and at steady state it becomes an algebraic equation. Rewrite Equation 2.10 as

$$a_{in} + \bar{t}R_A = a_{out} \tag{2.11}$$

where

$$\bar{t} = V/Q \tag{2.12}$$

is known as the **mean residence time**. We solve Equation 2.11 for four cases.

**EXAMPLE 2.1** Determine the steady state performance of a stirred tank reactor for each of the reactions in Equations 2.6 – 2.9.

**Solution:** We start with Equation 2.11 in all cases and substitute the appropriate value for $R_A$. Note that the reaction rate, $R_A$, is evaluated using the outlet concentration, $a_{out}$, since, by assumption for a stirred tank, the concentration inside the reactor is everywhere the same and is equal to $a_{out}$. After substitution for $R_A$, the resulting equation is solved for the **fraction unreacted**, $a_{out}/a_{in}$.

**Inert Tracer,** $R_A$ is given by Equation 2.6.

Equation 2.11 becomes

$$a_{in} = a_{out}$$

The fraction unreacted is

$$\frac{a_{out}}{a_{in}} = 1 \tag{o}$$

**First Order Reaction,** $R_A$ is given by Equation 2.7.

Equation 2.11 becomes

$$a_{in} = \bar{t}ka_{out} + a_{out}$$

The fraction unreacted is

$$\frac{a_{out}}{a_{in}} = \frac{1}{1 + k\bar{t}} \qquad \text{(i)}$$

Note that the product, $k\bar{t}$, is dimensionless. Steady state reactor performance, as measured by the dimensionless response $a_{out}/a_{in}$, depends only on $k\bar{t}$.

**Second Order, One Reactant,** $R_A$, is given by Equation 2.8.

Equation 2.11 becomes

$$a_{in} - 2\bar{t}k\,a_{out}^2 = a_{out}$$

The fraction unreacted is

$$\frac{a_{out}}{a_{in}} = \frac{-1 + \sqrt{1 + 8a_{in}k\bar{t}}}{4a_{in}k\bar{t}} \qquad \text{(ii)}$$

Note that the product, $a_{in}k\bar{t}$, is dimensionless. Steady state reactor performance is governed solely by this one dimensionless number.

**Second Order, Two Reactants,** $R_A$, is given by Equation 2.9.

This case requires two versions of Equation 2.11, one for each component. Note that $R_A = R_B$ since A and B are consumed at equal rates. The equations are

$$a_{in} - \bar{t}ka_{out}b_{out} = a_{out}$$

$$b_{in} - \bar{t}ka_{out}b_{out} = b_{out}$$

The solution is

$$\frac{a_{out}}{a_{in}} = \frac{-1-(b_{in}-a_{in})k\bar{t}+\sqrt{(1+b_{in}k\bar{t}-a_{in}k\bar{t})^2-4a_{in}k\bar{t}}}{2a_{in}k\bar{t}} \qquad \text{(iii)}$$

and

$$\frac{b_{out}}{b_{in}} = 1 + \frac{a_{in}}{b_{in}}\left(\frac{a_{out}}{a_{in}} - 1\right) \qquad \text{(iv)}$$

There are now two response variables: $b_{out}/b_{in}$ and $a_{out}/a_{in}$. Values for the response variables are determined by the two input variables, $a_{in}k\bar{t}$ and $a_{in}/b_{in}$.

---

## 2.3 BATCH REACTORS

We now consider transient operation of a stirred tank reactor. Such transient operation can be done in a variety of ways, the simplest of which is a batch reaction.

### 2.3.1 Single Batch Reactions

We suppose a reactor is initially empty but is filled very quickly. Inflow then stops and stays at zero throughout the remainder of the batch cycle. Outflow is zero until the very end of the batch cycle, at which time the vessel is quickly discharged and the cycle begins anew. Our job is to calculate a(t) during the filled portion of the batch cycle when V is constant and $Q_{in} = Q_{out} = 0$. Equation 2.5 gives

$$VR_A\ (a_{out}, b_{out}, ...) = \frac{d(a_{out} V)}{dt} \qquad (2.13)$$

Since V is constant, it can be eliminated. Also, the notation a,b,... for the various concentrations in the vessel is preferred over $a_{out}$, $b_{out}$, ... for this special case where there is no outlet flow. Thus,

$$\frac{da}{dt} = R_A \qquad (2.14)$$

with initial condition $a = a_0$ at $t = 0$. A similar equation can be written for each chemical component, thus giving a set of simultaneous first order ODEs as in Equation 1.1.

The batch reactor equation is easily solved for the four example reactions of Equations 2.6 – 2.9.

**EXAMPLE 2.2** Determine the fraction unreacted in a batch reactor at time t for each of the reactions in Equations 2.6 – 2.9.

**Solution:** We start with Equation 2.14 and substitute the appropriate function for $R_A$. As in Example 2.1, concentrations are constant throughout the reactor so that there is no spatial variation. There is, however, a temporal variation so that Equation 2.14 is a differential equation that must be integrated with respect to time. The initial condition in all cases is $a = a_o$ at $t = 0$.

**Inert Tracer,** $R_A$ is given by Equation 2.6.

Equation 2.14 becomes

$$\frac{da}{dt} = 0$$

The integral is

$$a = C$$

where C is a constant of integration.

Substituting the initial condition gives the solution

$$a = a_o$$

so that

$$\frac{a}{a_o} = 1 \qquad \text{(o)}$$

**First Order Reaction,** $R_A$, is given by Equation 2.7.

Equation 2.14 becomes

$$\frac{da}{dt} = -ka$$

The integral is

$$a = Ce^{-kt}$$

Applying the initial condition gives

$$a = a_o e^{-kt}$$

and

$$\frac{a}{a_o} = e^{-kt} \qquad \text{(i)}$$

Note that the product kt is dimensionless.

**Second Order, One Reactant,** $R_A$, is given by Equation 2.8.

Equation 2.11 becomes

$$\frac{da}{dt} = -2ka^2$$

Integration gives

$$-\frac{1}{a} = -2kt + C$$

Using the initial condition to evaluate the constant gives

$$a/a_o = \frac{1}{1 + 2ka_o t} \qquad \text{(ii)}$$

Note that k has units of [volume]/[mole]/[time] for a second order reaction so that $ka_ot$ is dimensionless.

**Second Order, Two Reactants,** $R_A$, is given by Equation 2.9.

This case requires two versions of Equation 2.11. They are

$$\frac{da}{dt} = -kab, \frac{db}{dt} = -kab$$

from which we obtain

$$\frac{da}{dt} = \frac{db}{dt}$$

Integration and application of the two initial conditions gives

$$a = a_o + b - b_o$$

Thus,

$$\frac{b}{b_o} = 1 + \frac{a_o}{b_o}\left(\frac{a}{a_o} - 1\right) \qquad \text{(iii)}$$

To find $a/a_o$, we eliminate b from the equation for da/dt:

$$\frac{da}{dt} = R_A = -kab = -k[a^2 + a(b_o - a_o)]$$

Integration and substitution of the initial condition gives

$$a/a_o = \frac{b_o - a_o}{b_o \exp[(b_o - a_o)kt] - a_o} \qquad \text{(iv)}$$

---

The reader should contrast the results in Example 2.2 with those in Example 2.1. Example 2.2 applies to a **batch** system. The reactor contents have **initial concentration** $a_o$; they are reacted for a **batch reaction time** t and have **final concentration** a(t). The dimensionless groups associated with these batch reactions are kt (first order reaction), $a_o kt$ (second order reaction), and $a_o/b_o$ (second order reaction of two reactants).

Example 2.1 applies to a special type of **flow** reactor: a perfectly mixed stirred tank reactor operated at steady state. The **inlet concentration** is $a_{in}$; the reactor has a **mean residence time** of $\bar{t}$; and the **outlet concentration** is $a_{out}$. The dimensionless groups associated with these flow reactions are $k\bar{t}$ (first order reaction), $a_o k\bar{t}$ (second order reaction), and $a_o/b_o$ (second order reaction of two reactants). For the same numerical values of these groups, the fractions unreacted will be higher (less reaction) in the stirred tank flow system than in the batch reactor. This is true for most kinds of flow systems and most types of reaction.

## 2.3.2 Multiple Reactions in Batch Reactors

All the examples in the previous section considered only a single chemical reaction. In fact, attention was limited to three specific cases of a single reaction.

$$A \rightarrow \text{Products}$$

$$2A \rightarrow \text{Products}$$

$$A + B \rightarrow \text{Products}$$

More complex reactions may exist, but the above are sufficient for most situations.

We consider now combinations of reactions, each member of which has one of the above forms. A simple combination is the following set of consecutive, first order reactions:

$$A \xrightarrow{k_I} B$$

$$B \xrightarrow{k_{II}} C$$

A key to understanding these reactions is to recognize that component B is formed by the first reaction but is consumed by the second. Thus the governing equations are

$$\frac{da}{dt} = -k_I a \tag{2.16}$$

$$\frac{db}{dt} = +k_I a - k_{II} b \tag{2.17}$$

$$\frac{dc}{dt} = +k_{II} b \tag{2.18}$$

The initial conditions are that $a = a_o$, $b = b_o$, and $c = c_o$ at $t = 0$. However, it is frequently true that $b_o = c_o = 0$ for such sets of consecutive reactions.

An analytical solution to Equations 2.16 – 2.18 can be found by integrating them sequentially. Equation 2.16 is integrated to give a(t). This result is then inserted into Equation 2.17 so that db/dt depends only on b and t. Equation 2.17 becomes an ODE with a single dependent variable, b. It too can be integrated, although not by the simple separation of variable technique we have used thus far for analytical solutions to single ODEs. We will postpone the analytical solution of Equations 2.16 – 2.18 until Chapter 3. A numerical solution can be found using the marching ahead technique introduced in Chapter 1.

**EXAMPLE 2.3** Solve Equations 2.16 – 2.18 numerically given the initial conditions $a = a_o$, $b = 0$, and $c = 0$ at $t = 0$.

**Solution:** We first scale the governing equations using the dimensionless variables

$$\alpha = a/a_o, \quad \beta = b/a_o, \quad \gamma = c/a_o, \quad \tau = k_I t$$

Note that we avoid scaling b by $b_o$ due to the initial condition that $b_o = 0$. Substituting the scaled variables into Equations 2.16 – 2.18 gives

$$\frac{d\alpha}{d\tau} = -\alpha \tag{i}$$

$$\frac{d\beta}{d\tau} = \alpha - \left(\frac{k_{II}}{k_I}\right)\beta \tag{ii}$$

$$\frac{d\gamma}{d\tau} = +\left(\frac{k_{II}}{k_I}\right)\beta \tag{iii}$$

The only visible parameter is $k_{II}/k_I$, but other parameters, $b_o/a_o$ and $\gamma_o/a_o$ are in the initial conditions.

The solution to (i) is just exp(-kτ). This could be substituted into (ii) to eliminate α. However, for numerical calculations it is essentially as easy to solve the entire set simultaneously.

A few results for the case $k_{II}/k_I = 1$, $\Delta\tau = .25$ are

| τ | α | β | γ |
|---|---|---|---|
| 0.25 | 0.750 | 0.250 | 0 |
| 0.50 | 0.563 | 0.375 | 0.063 |
| 0.75 | 0.422 | 0.422 | 0.156 |
| 1.00 | 0.316 | 0.422 | 0.262 |

These results are, of course, quite inaccurate due to the large value of Δτ.

The key programming steps in solving this set of equations are:

```
RA = -A
RB = +A - XK*B
RC = XK*B
A = A + DT*RA
B = B + DT*RB
C = C + DT*RC
T = T + DT
```

where XK denotes $k_{II}/k_I$. Everything else in the program is just initiation and print control.

Example 2.3 is fairly simple, as it involves only three components (A, B, C) and two reactions. Common situations such as fuel combustion or crude oil cracking can be much more complex, sometimes involving hundreds of components and reactions. The student of chemical reaction engineering will learn how to handle the general case of n components undergoing m reactions. This subject can be fairly complicated. We give a brief description for the sake of completeness and possible future reference. The student can skip directly to Section 2.4 without loss of continuity.

Consider some general reaction. We will call it Reaction I and denote it as

$$\nu_{A,I}A + \nu_{B,I}B + \nu_{C,I}C + \dots \rightarrow 0 \tag{2.19}$$

The $\nu$ represents stoichiometric coefficients. Thus $\nu_{A,I}$ is the stoichiometric coefficient of component A in Reaction I. If $\nu_{A,I} = 0$, then component A does not appear in Reaction I. If $\nu_{A,I}$ is negative, then component A is a **reactant** and is **consumed** by Reaction I. If $\nu_{A,I}$ is positive, then component A is a **product** and is **formed** by Reaction I. For Reaction II we write

$$\nu_{A,II}A + \nu_{B,II}B + \nu_{C,II}C \dots \rightarrow 0 \tag{2.20}$$

and so on Reactions III, IV, .... .

The rate of Reaction I is $R_{I.}$ The rate of formation of component A by Reaction I is

$$R_{A,I} = \nu_{A,I}R_I \tag{2.21}$$

Note that the sign of $\nu_{A,I}$ determines whether component A is formed or consumed by Reaction I. If component A is involved with several reactions, its overall rate of formation is

$$R_A = \sum_I R_{A,I} = \sum_I \nu_{A,I}R_I \tag{2.22}$$

where the summation extends over all reactions.

With these preliminaries aside, we are now able to treat the general case of multiple components reacting according to multiple reactions. For the case of a batch reactor, Equation 2.14 is written for each component, A, B, C, ... . For $R_A$, then, we use Equation 2.22 as the net rate of formation of component A (or B, C, ...) summed over all reactions. There will be one ODE for each component.

**EXAMPLE 2.4** Formulate the set of ODEs that governs the component concentrations in the following batch reaction scheme:

$$2A \rightarrow B, \qquad R_I = k_I a^2 \qquad (I)$$

$$A + B \rightarrow P, \qquad R_{II} = k_{II} ab \qquad (II)$$

$$A \rightarrow Q, \qquad R_{III} = k_{III} a \qquad (III)$$

**Solution:** The stoichiometric coefficients are

$$\nu_{A,I} = -2, \qquad \nu_{A,II} = -1, \qquad \nu_{A,III} = -1$$

$$\nu_{B,I} = +1, \qquad \nu_{B,II} = -1, \qquad \nu_{B,III} = 0$$

$$\nu_{P,I} = 0, \qquad \nu_{P,II} = +1, \qquad \nu_{P,III} = 0$$

$$\nu_{Q,I} = 0, \qquad \nu_{Q,II} = 0, \qquad \nu_{Q,III} = +1$$

The overall rates of formation for the various components are

$$R_A = R_{A,I} + R_{A,II} + R_{A,III}$$

$$= \nu_{A,I} R_I + \nu_{A,II} R_{II} + \nu_{A,III} R_{III}$$

$$= -2k_I a^2 - k_{II} ab - k_{III} a$$

$$R_B = k_I a^2 - k_{II} ab$$

$$R_P = + k_{II} ab$$

$$R_Q = + k_{III} a$$

The set of ODEs for a batch reaction is

$$\frac{da}{dt} = -2k_I a^2 - k_{II} ab - k_{III} a$$

$$\frac{db}{dt} = +k_I a^2 - k_{II} ab$$

$$\frac{dp}{dt} = + k_{II} ab$$

$$\frac{dq}{dt} = +k_{III} a$$

subject to the initial conditions that $a = a_o$, $b = b_o$, $p = p_o$, and $q = q_o$ at $t = 0$.

The general case of n components undergoing batch reaction in a stirred tank gives rise to a set of n, simultaneous, first order ODEs. This is the canonical problem of the book, and it can be solved using the marching ahead technique described in Chapter 1. The remainder of the book will deal with analytical solution techniques, other methods of numerical solutions, and other physical systems that give rise to sets of ODEs.

The overall rate of formation of a component, Equation 2.22, can also be used for multiple reactions in a steady state flow reactor. Equation 2.22 is just substituted into Equation 2.11, one version of which is written for each component. The result is a set of n, simultaneous, algebraic equations. Interestingly, it is often more difficult to solve the algebraic equations than the differential equations.

## 2.4 TRANSIENTS IN CONTINUOUS FLOW STIRRED TANKS

We continue the application of ODEs in stirred tank systems by considering cases with a net throughput, $Q > 0$.

### 2.4.1 Inert Tracer Washout

Suppose $Q_{in} = Q_{out} = Q$ and that a non-reactive tracer, $R_A = 0$, has been fed to the tank for a sufficient long time for steady state to have been established. At this steady state, $a_{out} = a_{in} = a_o$.

Now, at time $t = 0$, suppose the input of tracer is suddenly stopped: $a_{in} = 0$ for $t > 0$. We suppose the tracer concentration to be sufficiently low so that its discontinuation has no effect on Q. How does $a_{out}$ vary with time?

The material balance on component A for time $t > 0$ is found from Equation 2.5:

$$0 = Q\,a_{out} + \frac{d(Va_{out})}{dt} \tag{2.23}$$

where we have used the facts that $a_{in} = 0$ and $R_A = 0$. Since V is constant, the governing equation is

$$\left(\frac{V}{Q}\right)\frac{da_{out}}{dt} = a_{out} \tag{2.24}$$

subject to the initial condition that $a_{out} = a_o$ at $t = 0$. The solution is

$$\frac{a_{out}}{a_o} = e^{-Qt/V} \tag{2.25}$$

so that the tracer washes out of the reactor according to an exponential decay curve. The group Qt/V is dimensionless. The factor V/Q has units of time and is known as the **mean residence time.**

$$\bar{t} = V/Q \tag{2.26}$$

The mean residence time is an important property of a flow system and has a variety of applications in engineering analysis. Its reciprocal, Q/V, is known as the **dilution rate,** this term being common in biochemical engineering. In electrical engineering terminology, $\bar{t}$ would be known as the **time constant** for the system.

Equation 2.25 can be written as

$$\frac{a_{out}}{a_o} = e^{-t/\bar{t}} \tag{2.27}$$

The mean residence time is seen to be a natural scaling factor to reduce real time to dimensionless time.

Suppose that the tracer molecules, component A, have similar flow and mixing patterns as the overall fluid in the reactor. Then the fraction $a_{out}/a_o$ gives the **washout function** for the system:

$$w(t) = \text{fraction of fluid having a residence time greater than } t \tag{2.28}$$

where w(0) = 1, since all the fluid spends some time in the system, and w(∞) = 0, since all the fluid must eventually leave. For a stirred tank reactor, w(t) has the simple exponential form

$$w(t) = e^{-t/\bar{t}} \tag{2.29}$$

Other types of flow systems (piston flow reactors, two stirred tanks in series, and so on) will have different functions for w(t). The washout function gives the response of the system to a step change in inlet tracer concentration. This step change is of the negative or turn-off variety

$$a_{in} = a_o, \qquad t < 0$$

$$a_{in} = 0, \qquad t > 0$$

In scaled units, $\alpha = a_{in}/a_o$, this becomes a **unit** negative step change

$$\alpha = 1, \qquad t < 0$$

$$\alpha = 0, \qquad t > 0$$

The value of $\alpha$ exactly at $t = 0$ is unimportant for most forms of engineering analysis. When needed it is usually suitable to define it as

$$\alpha = 1/2 \qquad \text{at } t = 0$$

## 2.5 REACTOR STARTUPS AND SHUTDOWNS

The material balance equations for a stirred tank reactor can be used to analyze reactor transients including the filling and emptying steps. Here we give two examples to illustrate the general procedure of combining the overall fluid balance equation with a component balance equation.

**EXAMPLE 2.5** A first order reaction is occurring in a stirred tank reactor. The reactor is initially at steady-state; however, at time $t = 0$, shutdown begins by setting $Q_{in} = 0$ while leaving $Q_{out} = Q$. The reactor will go empty at time $\bar{t} =$ t. Determine $a_{out}$ for the period $0 < t < \bar{t}$.

Solution: At time $t = 0$, $a_{out}$ has its steady-state value given by Equation i of Example 2.1

$$a_{out} = \frac{a_{in}}{1 + k\bar{t}} \qquad \text{(i)}$$

For $t > 0$ the component balance on A gives

$$V(-k\, a_{out}) = Q\, a_{out} + \frac{d(V\, a_{out})}{dt} \qquad \text{(ii)}$$

where we have used Equation 2.5 with $Q_{in} = 0$. Here, V varies with time as the vessel is emptying. The overall balance equation gives

$$\frac{dV}{dt} = -Q$$

or

$$V = V_{full} - Qt$$

Substituting into ii gives

$$(V_{full} - Qt)(-ka_{out}) = Qa_{out} + (V_{full} - Qt)\frac{da_{out}}{dt} + a_{out}[-Q]$$

Many terms cancel to give the batch reaction equation for a first order reaction

$$-ka_{out} = \frac{da_{out}}{dt} \qquad \text{(iii)}$$

The initial condition is given by i. The solution is

$$a_{out}(t) = \frac{a_{in}\, e^{-kt}}{1 + k\bar{t}},\; 0 < t < \bar{t}$$

The continuous flow stirred tank behaves as a batch reactor when inflow is stopped. This is true whether or not outlet flow continues. Note that the extent of reaction, as measured by the **conversion**, $1\text{-}(a_{out}/a_{in})$, will overshoot its steady state concentration during a shutdown of this type.

**EXAMPLE 2.6** Analyze the shutdown transient for a continuous stirred tank reactor if the inlet reactant concentration, $a_{in}$, is suddenly stopped at time $t = 0$. For $t < 0$, A will be consumed by a first order reaction.

**Solution:** For this situation, Q and V are constant throughout the shutdown. Thus, only the component balance is needed. For $t > 0$, $a_{in} = 0$ and Equation 2.5 gives

$$V(-ka_{out}) = Qa_{out} + V\frac{da_{out}}{dt}$$

or

$$\frac{da_{out}}{dt} = -(k + Q/V)\, a_{out} = -\left(\frac{k\bar{t} + 1}{\bar{t}}\right) a_{out}$$

The initial condition is the original, steady state response.

$$a_{out} = \frac{a_{in}}{1 + k\bar{t}}$$

The solution is

$$\frac{a_{out}(t)}{a_{in}} = \frac{\exp\left[-\left(\frac{k\bar{t}+1}{\bar{t}}\right)t\right]}{1+k\bar{t}}$$

This result reduces to Equation 2.25 if k = 0. It is the washout function for a reactive tracer. The overall time constant,

$$\left(\frac{k\bar{t}+1}{\bar{t}}\right)^{-1} = \frac{\bar{t}}{1+k\bar{t}}$$

is smaller than that for a non-reactive tracer since A disappears both by washout and by reaction.

---

## PROBLEMS

2.1 Consider the following set of consecutive reactions

$$2A \rightarrow B$$

$$B \rightarrow C$$

The reaction rates are

$$R_A = -k_I a^2$$

$$R_B = +1/2\, k_I a^2 - k_{II} b$$

$$R_C = +k_{II} b$$

a) Explain the factor of 1/2 in the expression for $R_B$.

b) Transform the batch reactor equations to dimensionless form using $a_o$ to scale all concentrations and $(k_I a_o)^{-1}$ to scale time.

c) Solve the set using $k_I a_o = 1$, $k_I a_o / k_{II} = 1$. Optimize the batch reaction cycle to produce the maximum concentration of B. What is the maximum value of $\beta = b/a_o$ and at what value of $\tau = k_I a_o t$ does it occur?

2.2 Review Examples 2.5 and 2.6. Do you think it possible to devise a shutdown strategy that keeps constant conversion during the shutdown period? Devise one or else state why it is impossible.

2.3 Example 2.4 treated a set of three reactions involving four components in a batch reactor. Reformulate this example for a steady state stirred tank.

2.4 Formulate the set of ODEs that governs the following set of reactions occurring in a batch reactor

$$A \rightarrow C, \qquad R_I = k_I a$$

$$A + B \rightarrow P, \qquad R_{II} = k_{II} ab$$

$$B \rightarrow Q, \qquad R_{III} = k_{III} b$$

2.5 A variant of the second order, two reactant reaction is

$$A + B \rightarrow 2B, \quad R = kab$$

This is one example of an **autocatalytic** reaction where formation of the product, component B, increases the rate.

a) Determine the fraction of A which is unreacted in the outlet stream of a steady state stirred tank reactor.

b) Determine the fraction of A which is unreacted after batch reaction for time t.

c) Does this reaction follow the general rule that batch reactors have higher conversions than flow reactors. Why or why not?

d) Discuss the above solutions for the case of $b_{in} = 0$ and for the case of $b_o = 0$. How is steady state attained in the flow case?

# CHAPTER 3

# SOLUTION TECHNIQUES

This chapter describes solution techniques for a variety of ordinary differential equations. As indicated previously, an **ordinary** differential equation has only one independent variable, typically time, t, or a spatial coordinate, x. The independent variable in **dynamic** equations is, by definition, time. There can be several dependent variables, and in systems such as polymerizations, there is an infinite number of dependent variables. A **single** equation contains only one dependent variable. **Simultaneous** ODEs contain two or more dependent variables. To be solvable, there must be as many ODEs as there are dependent variables. Another requirement for solving a set of equations is that there be boundary or initial conditions. **Initial conditions** refer to dynamic equations and specify the values of the dependent variables, or their derivatives, at the initial time, $t = 0$. **Boundary conditions** specify values for the dependent variables, or their derivatives, at defined points in space. The term boundary condition is more general and includes initial conditions as a subcategory. To be solvable, a set of first order ODEs must contain as many boundary conditions as there are dependent variables. An $n^{th}$ order ODE requires n boundary conditions. Solvability of ODEs also depends on **linearity**. **Nonlinear** equations are rarely solvable analytically but can be solved numerically. Single, linear, first order ODEs are generally solvable. Higher order ODEs or sets of simultaneous, first order ODEs are generally solvable only when they have **constant coefficients**.

This chapter treats both analytical and numerical solutions of ODEs. We begin with analytical solutions to linear ODEs.

## 3.1 LINEARITY

The most general, linear, single, first order, ODE has the form

$$\frac{da}{dt} + P(t)a = Q(t) \tag{3.1}$$

where P and Q are arbitrary and typically nonlinear functions of t. A set of two simultaneous, linear, first order, ODEs has the form

$$\frac{da}{dt} + P_A(t)a + P_B(t)b = Q_A(t)$$

$$\frac{db}{dt} + S_A(t)a + S_B(t)b = Q_B(t) \qquad (3.2)$$

where $P_A$, $P_B$, $Q_A$, $S_A$, $S_B$, $Q_B$ are arbitrary and typically nonlinear functions of t.

The most general, linear, single $n^{th}$ order ODE has the form

$$\frac{d^n a}{dt^n} + P_{n-1}(t)\frac{d^{n-1}a}{dt^{n-1}} + \ldots + P_1(t)\frac{da}{dt} + P_o(t)a = Q(t) \qquad (3.3)$$

where $P_{n-1}$, $P_{n-2}$, ..., $P_1$, $P_0$, and Q are arbitrary and typically nonlinear functions of t. Note that the dependent variable and its derivatives appear as simple linear terms and not as products or powers. This is the essential feature of linearity.

The $n^{th}$ order, linear ODE with constant coefficients has the general form

$$\frac{d^n a}{dt^n} + A_{n-1}\frac{d^{n-1}a}{dt^{n-1}} + \ldots + A_1\frac{da}{dt} + A_0 a = Q(t) \qquad (3.4)$$

where $A_{n-1}$, ..., $A_0$ are constants and Q(t) is an arbitrary and typically nonlinear function of t.

In all the above examples, the term on the righthand side, Q(t), is known as the **forcing function**.

## 3.2 LINEAR, FIRST ORDER ODEs

The general form is given by Equation 3.1. This form has the solution

$$a = e^{-I(t)}\int Q(t)e^{I(t)}dt + Ce^{-I(t)} \qquad (3.5)$$

where C is a constant of integration and where

$$I(t) = \int P(t)dt \qquad (3.6)$$

All the integrals in Equations 3.5 and 3.6 are indefinite. Integration is performed to find a functional form, not a specific number. If the indefinite integrals cannot be evaluated analytically, then Equation 3.1 should be solved numerically.

**EXAMPLE 3.1** Solve

$$\frac{da}{dt} - \frac{2t}{1+t^2}\, a = 1$$

subject to the boundary condition that $a = 1$ at $t = 0$.

**Solution:** This has the form of Equation 3.1 with

$$P(t) = \frac{-2t}{1+t^2} \,.\quad Q(t) = 1$$

Equation 3.6 gives

$$I(t) = \int \frac{-2t}{1+t^2}\, dt = -\ln\,(1+t^2)$$

and thus

$$\exp[I(t)] = \frac{1}{1+t^2}\,,\quad \exp[-I(t)] = 1 + t^2$$

Substituting into Equation 3.5 gives

$$a = (1+t^2)\int \frac{1}{(1+t^2)}\, dt + C\,(1+t^2)$$

$$= (1+t^2)\,[\tan^{-1} t + C]$$

Applying the boundary condition gives C = 1. Thus, the solution is

$$a = (1 + t^2)\,[1 + \tan^{-1} t]$$

---

## 3.3 HOMOGENEOUS, LINEAR ODES WITH CONSTANT COEFFICIENTS

The general form is Equation 3.4. The equation is said to be **homogeneous** if Q(t) = 0.

The homogeneous equation is solved using the **characteristic equation**, where $d^n a/dt^n$ has been replaced by $s^n$:

$$s^n + A_{n-1}\, s^{n-1} + \dots + A_1 s + A_0 = 0 \qquad (3.7)$$

We treat this as a polynomial in s and solve for the roots $s_1, s_2, \dots, s_n$. If no roots are repeated, the general solution to the $n^{th}$ order, homogeneous ODE is

$$a = C_n e^{s_n t} + C_{n-1} e^{s_{n-1} t} + \dots + C_1 e^{s_1 t} \qquad (3.8)$$

If a root is repeated, then $s_i = s_j$ for some i and j. For repeated roots, the corresponding terms in Equation 3.8 are

$$(C_i + C_j t)\, e^{s_i t}$$

If $s_i = s_j = s_k$, the terms in Equation 3.8 are

$$(C_i + C_j t + C_k t^2)\, e^{s_i t}$$

and so on up to the level of repetition for each repeated root.

**EXAMPLE 3.2** Find a general solution to the following ODE:

$$\frac{d^2 a}{dt^2} - a = 0 \qquad (i)$$

**Solution:** The characteristic equation is

$$s^2 - 1 = 0$$

The roots are distinct:

$$s = \pm 1$$

Thus, the solution is

$$a = c_1 e^t + c_2 e^{-t}$$

as may be verified by differentiation. An equivalent solution to i uses hyperbolic functions $a = B_1 \sinh t + B_2 \cosh t$. This equivalent form arises whenever the characteristic equation has paired real roots, $s = \pm \sigma$.

---

**EXAMPLE 3.3** Find a general solution to

$$\frac{d^2a}{dx^2} - 4\frac{da}{dx} + 4a = 0$$

**Solution:** The characteristic equation is

$$s^2 - 4s + 4 = 0$$

The roots are $s_1 = 2$ and $s_2 = 2$ (repeated).

The solution is

$$a = C_1 e^{2x} + C_2 x e^{2x}$$

as may be verified by differentiation.

---

**EXAMPLE 3.4** Find a general solution to

$$\frac{d^2a}{dt^2} + 2\frac{da}{dt} + 2a = 0$$

**Solution:** The characteristic equation is

$$s^2 + 2s + 2 = 0$$

The roots are $s = -1 \pm i$ where $i = \sqrt{-1}$. (Some texts use electrical engineering notation where i is reserved for current and $j = \sqrt{-1}$. We prefer the mathematical/chemical engineering notation with $i = \sqrt{-1}$.) Thus,

$$a = C_1 e^{(-1+i)t} + C_2 e^{(-1-i)t}$$

Here the roots appear as a conjugate pair, $s = \sigma \pm \omega i$. The above expression using exponential functions with complex arguments is formally correct. However, a more useful form factors out the real portion of the exponential and replaces the imaginary portion with trigonometric functions:

$$a = e^{-t}(C_1 \sin t + C_2 \cos t)$$

That this is also a solution to the original ODE can be verified by differentiation.

---

An $n^{th}$ order linear ODE with constant coefficients can be solved by the above methods. The solution is an n-term series of exponential functions. Each term of the series has a multiplicative constant of integration. The arguments of the exponential functions are the roots of the **characteristic equation**, Equation 3.7. When roots are repeated, the repeated exponential terms are multiplied by t, $t^2$, and so on for each repetition. The characteristic equation has real coefficients so any imaginary roots must appear as conjugate pairs of complex numbers: $s = \sigma \pm \omega i$. In such cases, the exponential functions

$$C_1 e^{(\sigma+\omega i)t} + C_2 e^{(\sigma-\omega i)t}$$

can be replaced by

$$e^{\sigma t} [B_1 \sin \omega t + B_2 \cos \omega t]$$

The characteristic equation sometimes has paired real roots of the form $s = \pm\sigma$. In this case the exponential terms

$$C_1 e^{\sigma t} + C_2 e^{-\sigma t}$$

can be replaced by

$$B_1 \sinh \sigma t + B_2 \cosh \sigma t$$

Paired roots such as $s = \sigma \pm \omega i$ and $s = \pm\sigma$ can, of course, be repeated. The corresponding terms in the solution are then multiplied by t, $t^2$, and so on up to the level of repetition.

## 3.4 NONHOMOGENEOUS, LINEAR ODES WITH CONSTANT COEFFICIENTS

We now consider Equation 3.4 with $Q(t) \neq 0$. The general solution to this equation can be written as

$$a = a_c + a_p \tag{3.9}$$

where $a_c$ is the characteristic or complementary solution to the homogeneous solution with $Q(t) = 0$ and $a_p$ is a particular solution to the full, nonhomogeneous equation. The characteristic solution is found by the characteristic equation method of Section 3.2. The particular solution can be found for many Q(t) by using the **method of undetermined coefficients**. In this method, we determine the coefficients of a trial function, P(t), in order to satisfy the nonhomogeneous equation. Table 3.1 gives the appropriate trial functions for a variety of forcing functions.

**TABLE 3.1 - Undetermined Coefficients**

| FORCING FUNCTION | TRIAL FUNCTION |
|---|---|
| A (a constant) | B (a constant) |
| $Ae^{\alpha t}$ | $Be^{\alpha t}$ |
| $A\cos \alpha t$ or $A\sin \alpha t$ | $B_1\cos \alpha t + B_2\sin \alpha t$ |
| $A\cosh \alpha t$ or $A\sinh \alpha t$ | $B_1\cosh \alpha t + B_2\sinh \alpha t$ |
| $A\,t^n$ | $B_nt^n + B_1t^{n-1} + ... + Bo$ |
| $A\,t^ne^{\alpha t}\cos \beta t$ or $A\,t^ne^{\alpha t}\sin \beta t$ | $(B_nt^n + ... + B_o)e^{\alpha t}\cos \beta t + (C_nt^n + ... + C_o)\, e^{\alpha t}\sin \beta t$ |
| $A\,t^ne^{\alpha t}\cosh \beta t$ or $A\,t^ne^{\alpha t}\sin \beta t$ | $(B_nt^n + ... + B_o)e^{\alpha t}\cosh \beta t + (C_nt^n + ... + C_o)e^{\alpha t}\sinh \beta t$ |

The method consists of substituting the trial function into Equation 3.4 and determining the necessary values for the coefficients so that the trial function is indeed a solution. The coefficients are the terms B, $B_1$, ..., C, $C_1$, ..., in Table 3.1. They are not constants of integration to be determined from the boundary conditions. Instead, the coefficients are determined from the nonhomogeneous equation itself.

**EXAMPLE 3.5** Find a complete solution to

$$\frac{d^2a}{dt^2} - a = 1 \tag{i}$$

**Solution:** From Example 3.2, the characteristic solution is

$$a_c = C_1e^t + C_2e^{-t}$$

Table 3.1 gives a trial function of B (a constant) for the forcing function of Q(t) = 1. Substituting a = B into Equation i gives

$$-B = 1$$

so that

$$a_p = -1$$

Thus, the complete solution is

$$a = a_c + a_p = C_1e^t + C_2e^{-t} - 1$$

as may be verified by differentiation. Note that $C_1$ and $C_2$ are arbitrary constants that must be determined from the boundary conditions.

---

In some cases, the trial functions may be identical to terms in the characteristic solution. The trial function should then be multiplied by the lowest positive integral power of t as needed to eliminate all duplication.

If the forcing function is the sum of several terms, then the appropriate trial functions must also be summed.

Methods exist for finding particular solutions for forcing functions other than those in Table 3.1 Analytical solutions can also be found for linear equations with variable coefficients (e.g., Equation 3.3) and even for some nonlinear equations. Such solutions are beyond the scope of this work. If necessary, we will resort to numerical solutions.

**EXAMPLE 3.6** Find a complete solution to

$$\frac{d^2a}{dt^2} - 2\frac{da}{dt} + a = te^t - e^t \tag{i}$$

The characteristic equation has repeated roots, $s_1 = s_2 = +1$, so that

$$a_c = C_1 e^t + C_2 te^t$$

The forcing function is the sum of two terms, $te^t$ and $-e^t$, so that we would normally require two trial functions. However, the trial function for $te^t$ is

$$(B_1 t + B_0)e^t \tag{ii}$$

and this function includes the trial function needed for the $-e^t$ portion of the forcing function. Thus we need only one trial function, but the trial function of (ii) duplicates both terms of the characteristic solution. It must be multiplied by $t^2$ to eliminate all duplication. Thus, the trial function becomes

$$(Bt^3 + B_0 t^2)e^t \tag{iii}$$

Substitution into (i) shows that the trial function indeed satisfies the nonhomogeneous equation provided $B_1 = 1/6$ and $B_0 = -1/2$. The complete solution is

$$a = \left(C_1 + C_2 t - \frac{t^2}{2} + \frac{t^3}{6}\right)e^t$$

where $C_1$ and $C_2$ are arbitrary constants.

---

**EXAMPLE 3.7** Solve Equations 2.16 – 2.18 analytically to find a(t), b(t), and c(t) for the set of consecutive, first order, batch reactions:

$$A \xrightarrow{k_I} B \xrightarrow{k_{II}} C$$

**Solution:** See Example 2.3. The scaled equations are

$$\frac{d\alpha}{d\tau} = -\alpha \tag{i}$$

$$\frac{d\beta}{d\tau} = \alpha - K\beta \tag{ii}$$

$$\frac{d\gamma}{d\tau} = K\beta \tag{iii}$$

subject to $\alpha = 1$, $\beta = \gamma = 0$ at $\tau = 0$ and where $K = k_{II}/k_I$.

This set can be solved by successive substitution; first solving Equation i, substituting into Equation ii and so on. Thus,

$$\alpha = e^{-\tau}$$

and Equation ii becomes

$$\frac{d\beta}{d\tau} + K\beta = e^{-\tau} \qquad \text{(iv)}$$

The characteristic solution is

$$\beta_c = C_1 e^{-K\tau}$$

The particular function depends on the value of K. For $K \neq 1$, the trial function is $B\exp(-\tau)$ and there is no duplication of terms between it and the characteristic solution. Substitution of the trial function into Equation iv gives

$$B = \frac{1}{1-K}$$

The general solution for $\beta$ is thus

$$\beta = C_1 e^{-K\tau} - \frac{e^{-\tau}}{1-K}$$

Applying the initial condition that $\beta = 0$ at $\tau = 0$ gives

$$\beta = \frac{e^{-K\tau} - e^{-\tau}}{1-K} \qquad \text{(v)}$$

This is the general solution for $\beta = b/a_0$ for consecutive, first order reactions with unequal rate constants, $k_I \neq k_{II}$. Equation v presents an obvious problem when the two rate constants are exactly equal since this gives $K = 1$. In this special case the trial function duplicates a term in the characteristic function and must be multiplied by $\tau$. The characteristic solution for $K = 1$ is

$$\beta_c = C_1 e^{-\tau}$$

and the trial function is $B\tau\exp(-\tau)$. The general solution is

$$\beta = (C_1 + \tau)e^{-\tau}$$

Applying the initial condition gives

$$\beta = \tau e^{-\tau} \qquad \text{(vi)}$$

as a special solution for the case of K = 1.

The reader may wonder why β has an apparent discontinuity at $K = k_{II}/k_I = 1$. In fact, the discontinuity is only apparent, and β is a continuous function of K. Consider

$$\lim_{K\to 1} \beta = \lim_{K\to 1} \frac{e^{-K\tau} - e^{-\tau}}{1-K}$$

This limit has the 0/0 form needed to apply L'Hospital's rule. Differentiating both numerator and denominator with respect to K gives

$$\lim_{K\to 1} \beta = \lim_{K\to 1} \frac{-\tau e^{-K\tau}}{-1} = \tau e^{-\tau}$$

so that Equation v agrees with Equation vi in the limit. Computationally, Equation v still causes a problem near K = 1. The strictly numerical solution in Example 2.1 did not have this awkwardness.

---

From a computational viewpoint, there are many examples where a numerical solution is easier to evaluate than an analytical solution. However, the usefulness of a solution does not rest solely on its computational ease. Analytical solutions allow greater insight and have more utility in subsequent analyses than strictly numerical solutions.

## 3.5 REDUCTION OF A SET TO A SINGLE ODE

In Chapter 1 we stated that it is sometimes possible to reduce a set of simultaneous ODEs to a single, higher order ODE. Such reductions are rarely desirable for numerical solutions. Indeed, we usually go the other way and expand higher order equations to obtain equivalent sets of first order ODEs. However, reduction can be useful in obtaining analytical solutions.

**EXAMPLE 3.8** Solve the following set of ODEs.

$$\frac{da}{dt} + a + 2b = 2e^{-t} \qquad \text{(i)}$$

$$\frac{db}{dt} - a + 3b = 5 \qquad \text{(ii)}$$

subject to the initial conditions that a = b = 0 at t = 0.

**Solution:** This set of two, first order ODEs can be reduced to a single, second order ODE in either of the dependent variables, a or b. Suppose we choose to eliminate a. Then Equation ii is solved for a to give

$$a = \frac{db}{dt} + 3b - 5 \qquad \text{(iii)}$$

Differentiating,

$$\frac{da}{dt} = \frac{d^2b}{dt^2} + 3\frac{db}{dt} \qquad \text{(iv)}$$

Substituting Equations iii and iv into Equation i gives

$$\frac{d^2b}{dt^2} + 3\frac{db}{dt} + \frac{db}{dt} + 3b - 5 + 2b = 2e^{-\tau}$$

or

$$\frac{d^2b}{dt^2} + 4\frac{db}{dt} + 5b = 5 + 2e^{-\tau} \qquad \text{(v)}$$

The characteristic equation is

$$s^2 + 4s + 5 = 0$$

with roots

$$s = \frac{-4 \pm \sqrt{16-20}}{2} = -2 \pm i$$

We take the characteristic solution as

$$b_c = e^{-2t}\,[C_1 \sin t + C_2 \cos t\,]$$

The trial solution for the forcing function in Equation v is $B_1 + B_2 \exp(-t)$. A general solution for b is

$$b = e^{-2t}\,[C_1 \sin t + C_2 \cos t] + 1 + e^{-t}$$

where the constants $C_1$ and $C_2$ must be found from the boundary conditions. We directly know only one boundary condition on b: b = 0 at t = 0. A second boundary condition can be deduced from Equation ii. We know a = 0 at t = 0. Then Equation ii gives db/dt = 5 at $\tau = 0$, which constitutes a second boundary condition on b. Applying these two boundary conditions gives

$$b = 1 + e^{-t} + 2e^{-2t}\,[\sin t - \cos t]$$

Substitution back into Equation ii gives

$$a = -2 + 2e^{-t} + 4e^{-2t} \sin t$$

This example is somewhat misleading with respect to algebraic complexity. In typical cases, particularly with orders higher than 2, the algebra can become tedious indeed.

---

The reduction of a set of first order ODEs to a single, higher order ODE requires the elimination of all but one of the dependent variables. To eliminate a variable, say a, we must solve for it in the form

$$a = f\left( b,\ c\ ,\ldots,\ \frac{db}{dt},\ \frac{dc}{dt},\ \ldots,\ \frac{d^2b}{dt^2},\ \frac{d^2c}{dt^2},\ \ldots \right) \tag{3.10}$$

The important feature of Equation 3.10 is that the righthand side does not contain a or its derivatives. If the derivatives of a are needed, they are found by differentiation of Equation 3.10. Then a and its derivatives can be eliminated from the remaining members of the simultaneous set of ODEs. As a general rule, the original set of ODEs must be linear for this procedure to succeed. Otherwise, we may not be able to find explicit solutions in the form of Equation 3.10. The procedure works with linear equations having variable coefficients. However, as a general rule, the higher order ODE will also have variable coefficients and will be difficult to solve. Even second order ODEs with variable coefficients will rarely have solutions expressible as a finite sum of exponential functions. Their solutions typically involve Bessel functions, hypergeometric functions, Legendre polynomials, and other of the so-called higher functions of mathematical physics.

## 3.6 LINEARIZATION

In some situations it is desirable to linearize a set of ODEs. The analysis of process stability using linear control theory is one example. A goal of linearization is to find an approximate equation which represents the true solution reasonably well and which is analytically tractable.

Linearization is done about a point in the space of dependent variables. Suppose there is only one dependent variable, y, and that its differential equation is

$$\frac{dy}{dt} = f(y) \tag{3.11}$$

We can linearize Equation 3.11 about the point $y_s$ by expanding f in a Taylor series about this point.

$$f(y) = f(y_s) + \frac{df}{dy}\Big|_{y_s}(y-y_s) + \frac{d^2f}{dy^2}\Big|_{y_s}\frac{(y-y_s)^2}{2!} + \dots \tag{3.12}$$

We assume that the value of $y_s$ is known and note that the various coefficients in the Taylor series

$$f(y_s),\ \frac{df}{dy}\Big|_{y_s},\ \frac{d^2f}{dy^2}\Big|_{y_s},\ \dots$$

are constants that can be calculated.

If we retain only the zero order term in the expansion, the ODE becomes

$$\frac{dy}{dt} \approx f(y_s) \tag{3.13}$$

The initial condition associated with Equation 3.13 is that $y = y_s$ at $t = t_s$. The solution is

$$y = y_s + (t - t_s)\,f(y_s) \tag{3.14}$$

This is a simple enough expression, and, if $t - t_s$ is small, it is even a good approximation to the rigorous solution of Equation 3.11. However, we can find a better approximation without too much more work. Retain the zero and first order terms so that the ODE becomes

$$\frac{dy}{dt} \approx f(y_s) + \frac{df}{dy}\Big|_{y_s}(y-y_s) = \left[f(y_s) - \frac{df}{dy}\Big|_{y_s}y_s\right] + \frac{df}{dy}\Big|_{y_s}y \tag{3.15}$$

This has the same initial condition: $y = y_s$ at $t = t_s$. The functional form of 3.15 is

$$\frac{dy}{dt} \approx A + By \tag{3.16}$$

which is readily integrable. The solution is

$$y = \frac{(By_s + A)\, e^{B(t-t_s)} - A}{B}$$

$$= \frac{f(y_s)\, e^{\left(\frac{df}{dy}\big|_{y_s}\right)(t-t_s)} - \left[f(y_s) - \frac{df}{dy}\big|_{y_s} y_s\right]}{\frac{df}{dy}\big|_{y_s}} \tag{3.17}$$

We can continue this procedure one more step to obtain a second order approximation in the form

$$\frac{dy}{dt} \approx A + By + Cy^2 \tag{3.18}$$

This has the same boundary condition and remains analytically integrable. However, the functional form of the integral now depends on the sign of $4AC-B^2$. Zero and second order approximations have specialized applications, but the first order or linear approximation, Equation 3.17, is the most useful. This is particularly true in control applications where linearization is done about a steady state point for which $f(y_s) = 0$. Then the zero order approximation predicts y will never change from $y_s$, but the first order approximation gives useful information regarding stability.

**EXAMPLE 3.9** Approximate $\frac{dy}{dt} = -y^2$ at the point $y_s = 1$ using a Taylor series expansion for $-y^2$. Set $t_s = 0$.

Solution: Here, $f(y) = -y^2$ giving $f(y_s) = f(1) = -1$; $\frac{df}{dy}\big|_{y_s} = -2y_s = -2$;

$$\frac{d^2f}{dy_s^2} = -2.$$

The zero order approximation is

$$y \approx 1 - t$$

The first order approximation is

$$y = \frac{-e^{-2t} - (-1+2)}{-2} = \frac{1}{2} + \frac{1}{2}e^{-2t}$$

The second order approximation gives

$$\frac{dy}{dt} = -1 - 2(y-1) = 2(y-1)^2$$

$$= -y^2$$

which is, of course, exact. The exact solution is

$$y = \frac{1}{1+t}$$

Some comparative results are

| t | Zero Order | First Order | Exact |
|---|---|---|---|
| 0 | 1 | 1 | 1 |
| 0.01 | 0.9900 | 0.9901 | 0.9901 |
| 0.1 | 0.9000 | 0.9094 | 0.9091 |
| 0.5 | 0.5000 | 0.6839 | 0.6667 |

---

The above example began with Equation 3.11 where f(y) has no explicit dependence on t. Linearization can also be used where f = f(y,t). Now, however, the resulting, linear ODE will have variable coefficients. The first order case can be integrated using the method of Section 3.2.

**EXAMPLE 3.10** Linearize

$$\frac{da}{dt} = \frac{t}{a} \qquad \text{(i)}$$

about its initial point, a = 1 at t = 0. Solve the resulting linear equation and compare this approximate solution to the exact solution.

**Solution:** f(a,t) = t/a. We expand this function in a Taylor series about the point a = 1 to give

$$f(a,t) = f(1,t) + \frac{\partial f}{\partial a}\Big|_1 (a-1) + \ldots$$

$$= t + (-t)(a-1) + \ldots$$

Neglecting terms of higher than first order gives

$$\frac{da}{dt} + ta = 2t$$

which has the form

$$\frac{da}{dt} + P(t)a = Q(t)$$

The method of Section 3.2 gives

$$a = 2 + Ce^{-t^2/2}$$

Applying the initial condition gives

$$a = 2 - e^{-t^2/2} \qquad \text{(ii)}$$

This result is to be compared to the exact solution which can be obtained by noting that Equation i is variable separable. The exact solution is

$$a = \sqrt{1+t^2} \qquad \text{(iii)}$$

Some comparative results are

| t | Linear Approximation Equation (ii) | Exact Solution Equation (iii) |
|---|---|---|
| 0 | 1 | 1 |
| 0.1 | 1.0050 | 1.0050 |
| 0.5 | 1.1175 | 1.1180 |
| 1 | 1.3935 | 1.4142 |

Consider now a set of simultaneous, nonlinear ODEs:

$$\frac{da}{dt} = R_A\,(a,b,\ldots)$$

$$\frac{db}{dt} = R_B\,(a,b,\ldots) \tag{3.19}$$

$$\vdots \qquad \vdots$$

Suppose there are N dependent variables. We use an N-dimensional Taylor series to linearize about the point $a_s$, $b_s$, ... . Dropping terms higher than first order gives

$$\frac{da}{dt} \approx R_A\,(a_s, b_s, \ldots) + \frac{\partial R_A}{\partial a}\Big|_{a_s} (a-a_s) + \frac{\partial R_A}{\partial b}\Big|_{b_s} (b-b_s) + \ldots$$

$$\frac{db}{dt} \approx R_B\,(a_s, b_s, \ldots) + \frac{\partial R_B}{\partial a}\Big|_{a_s} (a-a_s) + \frac{\partial R_B}{\partial b}\Big|_{b_s} (b-b_s) + \ldots$$

These ODEs are linear. They will have constant coefficients if t does not explicitly appear in $R_A$, $R_B$, ... . If t does so appear, an analytical solution will be difficult if $N > 1$. Approximation techniques can be devised for equations with variable coefficients, but are not especially useful.

## 3.7 NUMERICAL SOLUTIONS

Chapter 1 described the Euler or marching ahead method for solving sets of ordinary differential equations. This method is extremely simple from a conceptual and programming viewpoint. It is computationally inefficient in the sense that a great many arithmetic operations are necessary to produce accurate solutions. More efficient techniques should be used when the same set of equations is to be solved many times, as in optimization studies. One such technique, **fourth order Runge-Kutta**, has proved quite popular and can be generally recommended for all but very stiff sets of first order, ordinary differential equations. (A set of ODEs is said to be stiff when the various time constants differ by orders of magnitude.)

The set of equations to be solved is

$$\frac{da}{dt} = R_A\,(a,b, \ldots t)$$

$$\frac{db}{dt} = R_B\,(a,b, \ldots t)$$

$$\vdots \qquad \vdots$$

A value of $\Delta t$ is selected, and values for $\Delta a$, $\Delta b$, ... are estimated by evaluating the functions $R_A$, $R_B$, ... . In the marching ahead method, this evaluation is done at the initial point $(a_0, b_0, \ldots, t_0)$ so that the estimate for $\Delta a$ is just $\Delta t R_A(a_0,b_0,\ldots,t_0) = \Delta t(R_A)_0$. In fourth order Runge-Kutta, the evaluation is done at four points and the estimates for $\Delta a$, $\Delta b$, ... are based on weighted averages of the $R_A$, $R_B$, ... at these four points:

$$\Delta a = \Delta t\,\frac{(R_A)_0 + 2(R_A)_1 + 2(R_A)_2 + (R_A)_3}{6}$$

$$\Delta b = \Delta t\,\frac{(R_B)_0 + 2(R_B)_1 + 2(R_B)_2 + (R_B)_3}{6}$$

$$\vdots \qquad \vdots$$

where

$$(R_A)_0 = R_A(a_0,b_0,\ldots, t_0)$$

$$(R_A)_1 = R_A(a_1,b_1,\ldots, t_1)$$

$$(R_A)_2 = R_A(a_2,b_2,\ldots, t_2)$$

$$(R_A)_3 = R_A(a_3,b_3,\ldots, t_3)$$

with similar equations applying for B,C, ... . The values for $a_1$, $a_2$, ... are determined in a sequential fashion:

$$a_1 = a_0 + \frac{1}{2}\Delta t\,(R_A)_0$$

$$a_2 = a_0 + \frac{1}{2}\Delta t\,(R_A)_1$$

$$a_3 = a_0 + \Delta t\,(R_A)_2$$

with similar equations holding for $b_1$, $b_2$, ... . Time rarely appears explicitly in the R, but should it appear,

$$t_1 = t_0 + \frac{1}{2}\Delta t$$

$$t_2 = t_1$$

$$t_3 = t_0 + \Delta t$$

---

**EXAMPLE 3.11** Solve the following batch reaction set for t = 1 given $a_0 = c_0 = 30$, $b_0 = p_0 = 0$, $k_I = 0.01$, and $k_{II} = 0.02$. Use fourth order Runge-Kutta integration with $\Delta t = 1.0$ and 0.5.

The set of chemical reactions is

$$2A \rightarrow B$$

$$B + C \rightarrow P$$

The corresponding set of ODEs is

$$\frac{da}{dt} = -2k_I a^2 \qquad \text{(i)}$$

$$\frac{db}{dt} = +k_I a^2 - k_{II} bc \qquad \text{(ii)}$$

$$\frac{dc}{dt} = -k_{II} bc \qquad \text{(iii)}$$

$$\frac{dp}{dt} = +k_{II} bc \qquad \text{(iv)}$$

Note the factor of 2 that appears in Equation i. Compare this set of ODEs to those in Problem 2.1. Either approach gives stoichiometrically correct results; however, observe that for the same chemical reaction the rate constant, $k_I$, in the current example is half that in Problem 2.1. The literature uses both approaches so that care is necessary in defining rate constants. See Section 2.3.2 or Nauman, **Chemical Reactor Design**, New York, Wiley, 1987, for a consistent method for treating stoichiometric coefficients and complex reaction networks.

**Solution:** For $\Delta t = 1.0$:

| i | $a_i$ | $b_i$ | $c_i$ | $(R_A)_i$ | $(R_B)_i$ | $(R_C)_i$ |
|---|---|---|---|---|---|---|
| 0 | 30.000 | 0 | 30.000 | -18.000 | 9.000 | 0 |
| 1 | 21.000 | 4.500 | 30.000 | -8.820 | 1.710 | -2.700 |
| 2 | 25.590 | 0.855 | 28.650 | -13.097 | 6.059 | -0.490 |
| 3 | 16.903 | 6.059 | 29.510 | -5.714 | -0.719 | -3.576 |
| Δ | -11.258 | 3.970 | -1.659 | | | |
| t=1 | 18.742 | 3.970 | 28.341 | | | |

For $\Delta t = 0.5$

| i | $a_i$ | $b_i$ | $c_i$ | $(R_A)_i$ | $(R_B)_i$ | $(R_C)_i$ |
|---|---|---|---|---|---|---|
| 0 | 30.000 | 0 | 30.000 | -18.000 | 9.000 | 0 |
| 1 | 25.500 | 2.250 | 30.000 | -13.005 | 5.153 | -1.350 |
| 2 | 26.749 | 1.288 | 29.663 | -14.310 | 6.391 | -0.764 |
| 3 | 22.845 | 3.195 | 29.236 | -10.438 | 3.351 | -1.868 |
| Δ | -6.992 | 2.953 | -0.508 | | | |
| t=1/2 | 23.078 | 2.953 | 29.492 | | | |
| 0 | 23.078 | 2.953 | 29.492 | | | |
| 1 | 20.415 | 3.849 | 29.057 | | | |
| 2 | 20.993 | 3.436 | 28.933 | | | |
| 3 | 18.670 | 4.162 | 28.498 | | | |
| Δ | -4.328 | 1.116 | -1.047 | | | |
| t=1 | 18.750 | 4.069 | 28.445 | | | |

---

Runge-Kutta is a powerful integration technique that remains fairly easy to implement. The fourth order Runge-Kutta algorithm converges $0(\Delta t^5)$. This means that the error between the approximate solution and the true solution decreases by a factor of 32 when the step size is halved. By comparison, the marching ahead technique converges $0(\Delta t)$ so that the error decreases by a factor of two when $\Delta t$ is halved. These statements are rigorously true only in the limit of $\Delta t \to 0$, but the powerful advantage of Runge-Kutta integration is apparent.

The Runge-Kutta algorithm will be unstable if the step size is too large. In a set of simultaneous equations, the algorithm uses the same step size for each member of the set. This obviously causes problems if the set is stiff. Most alternatives to Runge-Kutta suffer similar problems, but a variety of fairly complex, "canned" computer programs is now available for stiff sets. See M.E. Davis, **Numerical Methods and Modeling for Chemical Engineers**, New York, Wiley, 1984, for a summary and comparison of available routines.

## PROBLEMS

3.1 Solve

$$\frac{d^2a}{dt^2} + \beta a = e^{-t}$$

subject to the initial conditions that $a = 0$ and $\frac{da}{dt} = 0$ at $t = 0$.

3.2 Repeat Problem 3.1 for the case of $\beta = -1$.

3.3 Find a complete solution to the following equation:

$$\frac{d^2y}{dx^2} + 5\frac{dy}{dx} + 6y = 3e^{-2x} + e^{3x}$$

3.4 Find a general solution to

$$(2y + x^2)dx = x\,dy$$

3.5 The substitution $z = a^{1-n}$ will reduce the Bernoulli equation

$$\frac{da}{dt} + P(t)a = Q(t)a^n$$

to a linear, first order equation. Use this fact to find a general solution to

$$a\frac{da}{dt} + a^2 = t$$

3.6 Find a general solution to the following set of ODEs:

$$\frac{da}{dt} + 2a + \frac{db}{dt} + 6b = 2e^t$$

$$2\frac{da}{dt} + 3a + 3\frac{db}{dt} + 8b = -1$$

3.7 Linearize and solve the following ODE about the point $y_S = 1$:

$$\frac{dy}{dt} = -y^3$$

Set $t_s = 0$. Evaluate the linearized solution at $t = 0$, 0.01, 0.1, and 0.5 and compare to the exact solution. Why are the errors larger in this problem than in Example 3.8?

3.8 Linearize and solve the following ODE about its initial point, $a = 1$ at $t = 0$:

$$\frac{da}{dt} = \frac{t}{a^2}$$

3.9 Compare the Euler and Runge-Kutta algorithms for solving the following equation:

$$\frac{db}{dt} = -b^2 - be^{-2t}$$

The initial condition is $b = 1$ at $t = 0$. Continue integration until $t = 2$. Repeat your calculations for several step sizes.

# CHAPTER 4

# LAPLACE TRANSFORMS

This chapter introduces a classical technique in engineering analysis, that of Laplace transformation. Our treatment will be more or less standard.

The major use of Laplace transformation is to solve linear ODEs. In essence, this technique converts an ordinary differential equation into an algebraic equation. (In advanced applications, it converts a partial differential equation to an ordinary differential equation. A second transformation then converts the ODE to an algebraic equation.) The result is conceptually similar to the characteristic equation method of solving homogeneous, linear ODEs with constant coefficients. However, Laplace transforms can also treat nonhomogeneous equations and, occasionally, equations with variable coefficients within the same theoretical framework used for homogeneous ODEs.

## 4.1 THE TRANSFORMATION

Given a function, f(t), its Laplace transform is defined as

$$£[\, f(t)] = \int_0^\infty f(t)\, e^{-st} dt = F(s) \tag{4.1}$$

where s is the transform parameter and has units of $[\text{t-units}]^{-1}$. For dynamic problems, s has units of inverse time. The pound sign notation is cumbersome and rarely used. The usual notation for the transformation is lower case for the t-dependent function and upper case for the s-dependent function. Thus,

Equation 4.1 defines the transformation $f(t) \rightarrow F(s)$. The notation $f(t) \rightarrow \bar{f}(s)$ is also used for this transformation, but we reserve the overbar to represent an average.

For theoretical purposes, it is necessary to regard the transform parameter as a complex variable:

$$s = \sigma + i\omega \tag{4.2}$$

where $i = \sqrt{-1}$. The use of Equation 4.1 to transform most functions, f(t), and to determine many general properties of the transformation does not require this interpretation. We need only suppose that s is sufficiently large to ensure convergence of the integral in Equation 4.1.

## 4.2 TRANSFORMS OF FUNCTIONS

The functions that we ordinarily deal with are continuous with continuous derivatives throughout the interval $0 < t < \infty$. Integration of Equation 4.1 causes no special difficulty. Many results can be found using an ordinary table of indefinite integrals and then applying the upper and lower limits of Equation 4.1.

**EXAMPLE 4.1** Find the Laplace transform of $e^{at}$

**Solution:** We seek

$$F(s) = \int_0^\infty e^{+at}e^{-st}dt = \int_0^\infty e^{(a-s)t}\,dt$$

The indefinite integral of $e^{bt}$ is

$$\int e^{bt} = \frac{e^{bt}}{b}$$

Thus, we seek

$$F(s) = \frac{e^{(a-s)t}}{a-s}\Big|_0^\infty = \lim_{t\to\infty}\frac{e^{(a-s)t}}{a-s} - \lim_{t\to 0}\frac{e^{(a-s)t}}{a-s}$$

or

$$F(s) = \frac{-1}{a-s} = \frac{1}{s-a}$$

where we have assumed $s > a$ so that the upper limit, $t \rightarrow \infty$, exists.

**EXAMPLE 4.2** Find the Laplace transform of $e^{-at}$

**Solution:** Proceeding as in Example 3.1,

$$F(s) = \frac{1}{s+a}$$

where now we need only assume s > -a to assure convergence.

---

Functions such as $e^{-at}$ and sin (at) are **bounded** over the interval $0 < t < \infty$. Thus, Equation 4.1 will converge provided only that s > 0. A great many unbounded functions such as $t^n$, t sin(at), and sinh(at) will also converge provided s is sufficiently large. Table 4.1 (see end of chapter) provides a set of Laplace transform pairs. The major use of Table 4.1 is to find an inverse transform, f(t), given the Laplace transform, F(s). Such inversion problems arise in the solution of ODEs.

### 4.2.1 The Unit Step Function

We now consider functions which have a few isolated points of discontinuity in the interval $0 \le t < \infty$. Such functions are said to be **piecewise continuous**. The most important member of this class is the **unit step function**, defined as

$$\begin{aligned} U(t) &= 0, \quad t<0 \\ U(t) &= 1, \quad t>0 \end{aligned} \tag{4.3}$$

The unit step function has a finite discontinuity at t = 0. This finite discontinuity does not contribute to the integral in Equation 4.1. Thus,

$$\pounds[U(t)] = \int_0^t e^{-st}dt = \frac{1}{s} \tag{4.4}$$

The same integral is found for a function which is constant over the doubly infinite interval:

$$f(t) = 1, \quad -\infty < t < \infty$$

gives

$$\pounds[1] = 1/s \tag{4.5}$$

The fact that Equations 4.4 and 4.5 are identical is an inherent ambiguity of the Laplace transform. For most problems, we can resolve this ambiguity by assuming the system to be **initially relaxed**. This means that the various dependent variables, their derivatives, and their forcing functions are identically zero for $t < 0$.

### 4.2.2 The Unit Impulse Function

We introduce now a very special function of engineering analysis known as the unit impulse function, $\delta(t)$. It is also known as the Dirac delta function after a famous physicist or simply as the delta function due to its notation, $\delta(t)$. It can be defined as

$$\delta(t) = 0 \qquad t > 0 \tag{4.6}$$

$$\int_0^\infty \delta(t) = 1 \tag{4.7}$$

This curious definition has $\delta(t)$ equal to zero everywhere except at a single point, $t = 0$. Unlike the unit step change, the discontinuity in $\delta(0)$ is so large that $\delta(0)$ cannot have a finite value. Indeed, $\delta(0)$ is so large that the integral of $\delta(t)$ has unit value. We can think of $\delta(t)$ as being a very sharp pulse at $t = 0$. The width of the pulse is zero. The height of the pulse is infinity. The area under the pulse has the indeterminate form, zero times infinity, which we shall define in this case as 1.0.

**EXAMPLE 4.3** Find a conventional function, $D(t)$, which in a suitable limit behaves as $\delta(t)$.

**Solution:** There is no unique function, $D(t)$, but a great many functions that will behave as $\delta(t)$ in the limit as the pulse width goes to zero. One such function is the rectangular pulse

$$D(t) = 0, \qquad t < 0$$

$$D(t) = 1/\sigma, \qquad 0 < t < \sigma$$

$$D(t) = 0, \qquad t > \sigma$$

The pulse width is $\sigma$. The pulse height is $\sigma^{-1}$. The area under the pulse is $\sigma/\sigma = 1$.

---

Pure mathematicians disdain informally defined functions such as $\delta(t)$ and have invented theories of integration, such as Steiltjes integration, to avoid their use. Engineers and applied mathematicians like $\delta(t)$ because of its manipulative convenience. Most formal operations (meaning manipulations) using impulse functions give correct answers, which, if required, can be justified by a limiting process such as that in Example 4.3.

The formal properties of $\delta(t)$ include

$$\int_0^{\infty} f(t)\,\delta(t-t_0)dt = f(t_0) \tag{4.8}$$

which is known as the shifting property.

**EXAMPLE 4.4** Use D(t) as defined in Example 4.3 to verify Equation 4.8 by a limiting process.

**Solution:** We replace $\delta(t\text{-}t_0)$ with $D(t\text{-}t_0)$ in Equation 4.8 to give

$$\int_0^{\infty} f(t)D(t-t_0)dt = \int_{-t_0}^{\infty} f(\theta+t_0)D(\theta)d\theta = \int_0^{\infty} f(\theta+t_0)D(\theta)d\theta$$

where we have substituted $\theta = t\text{-}t_0$. Using the definition of D in Example 4.3

$$\int_0^{\infty} f(t)\delta(t-t_0)dt = \lim_{\sigma\to 0}\int_0^{\infty} f(\theta+t_0)D(\theta)d\theta = \lim_{\sigma\to 0}\int_0^{\sigma}\frac{f(\theta+t_0)}{\sigma}d\theta = \lim_{\sigma\to 0}\frac{\theta\bar{f}(\theta+t_0)}{\sigma}\Big]_0^{\sigma}$$

where $\bar{f}$ is some value of f(t) in the range $\theta < t < \theta + t_0$. Taking the limit as $\sigma \to 0$ gives

$$\int_0^{\infty} f(t)D(t-t_0)dt = \lim_{\sigma\to 0}\bar{f}(\sigma+t_0) = f(t_0)$$

---

The most important property of $\delta(t)$ is its Laplace transform.

$$\pounds[\delta(t)] = \int_0^{\infty}\delta(t)e^{-st}dt = 1 \tag{4.9}$$

This result follows directly from the shifting property, Equation 4.8, with $t_0 = 0$.

## 4.3 GENERAL TRANSFORM PROPERTIES

The previous section showed how to take the Laplace transform of various functions including U(t) and δ(t). We now consider properties of £[f(t)] which are valid for more general functions, f(t).

### 4.3.1 Derivatives

Suppose we know the Laplace transform, F(s), of the function f(t). We would like to know the transform of the derivatives of f(t). By definition,

$$F(s) = \int_0^\infty f(t)e^{-st}dt$$

We integrate by parts using

$$u = f(t) \qquad du = f'(t)dt = df$$
$$v = -\frac{1}{s}e^{-st} \qquad dv = e^{-st}dt$$

so that

$$F(s) = uv \Big|_0^\infty - \int_0^\infty v\,du = \frac{-f(t)}{s}e^{-st}\Big|_0^\infty + \int_0^\infty \frac{f'(t)e^{-st}}{s}dt \tag{4.10}$$

In writing these equations, we suppose that £[f(t)] = F(s) and £[f '(t)] exist. This means that the defining integrals, as in Equation 4.1, must converge. In turn, this implies that

$$\lim_{t\to\infty} f(t)e^{-st} = 0 \text{ and } \lim_{t\to\infty} f'(t)e^{-st} = 0 \tag{4.11}$$

Using the first of these limits, Equation 4.10 can be rewritten as

$$£\,[f'(t)] = \int_0^\infty f'(t)\,e^{-st}dt = sF(s) - f(0) \tag{4.12}$$

where f(0) denotes the initial value of f at $t = 0$. In Laplace transform theory, we treat initially relaxed functions: $f(t) = 0,\ t < 0$. A discontinuity may occur at $t = 0$ so that $f(t) \neq 0$ for $t > 0$. In Equation 4.12, the initial value f(0) means f(0+), after any discontinuity has occurred. Thus, Equation 4.12 is better written as

$$£[f'(t)] = sF(s) - f(0+) \tag{4.13}$$

where f(0+) denotes the initial value of f(t) after any possible discontinuity at t = 0.

The integration by parts technique can be repeated to determine £[F''(t)] = £[$d^2f/dt^2$] and so on. The results are

$$£[F''(t)] = £[d^2f/dt^2] = s^2 F(s) - sf(0+) - f'(0+) \tag{4.14}$$

and

$$£[f^{(n)}(t)] = £[d^nf/dt^n] = s^nF(s) - s^{n-1}f(0+) - s^{n-2}f'(0+) - ... - f^{(n)}(0+) \tag{4.15}$$

Laplace transformation of an nth derivative gives an nth order polynomial in s. The coefficients of the various terms of this polynomial depend on the initial values: f(0+), f' (0+), f'' (0+), ..., $f^{(n-1)}(0+)$. We shall see that the general solution to an nth order ODE with constant coefficients can be obtained by Laplace transformation techniques. If the various initial values, f(0+), f' (0+), ..., are all known, the Laplace transformation technique will yield a final solution with all constants evaluated. If some of the initial values are unknown, they will appear as arbitrary constants to be determined from other (noninitial) boundary conditions.

## 4.3.2 Linearity

The most important property of Laplace transformation is that of **linearity** whereby

$$£[C_1f(t) + C_2g(t)] = C_1F(s) + C_2G(s) \tag{4.16}$$

where $C_1$ and $C_2$ are constants.

This property can be confirmed from the definition of Laplace transformation, Equation 4.1:

$$£[C_1f(t) + C_2g(t)] = \int_0^\infty [C_1f(t) + C_2g(t)]\, e^{-st}dt$$

$$= C_1\int_0^\infty f(t)e^{-st}dt + C_2\int_0^\infty g(t)e^{-st}dt = C_1F(s) + C_2G(s)$$

This result shows that £ is a **linear operator.**

**EXAMPLE 4.5** Take the Laplace transform of

$$y'' - 6y' + 3y = e^{-t} \tag{i}$$

subject to the initial conditions $y'(0+) = y(0+) = 0$.

**Solution:** The Laplace transform is

$$\int_0^\infty (y'' - 6y' + 3y)e^{-st}\,dt = \int_0^\infty e^{-t}e^{-st}dt$$

This becomes

$$[s^2Y(s) - sy(0+) - y'(0+)] - 6[sY(s) - y(0+)] + 3[Y(s)] = \frac{1}{s+1} \qquad \text{(ii)}$$

where we have set $y = f(t)$ and used the differentiation results to evaluate the left-hand side of Equation (i). The right-hand side is found from $\pounds[e^{-t}]$ as given in Table 4.1. Applying the initial conditions gives

$$s^2Y(s) - 6sY(s) + 3Y(s) = \frac{1}{s+1}$$

or

$$Y(s) = \frac{1}{(s+1)(s^2 - 6s + 3)} \qquad \text{(iii)}$$

---

### 4.3.3 Uniqueness

Example 4.5 showed how an ODE, together with its initial conditions, can be reduced to an algebraic function, Y(s). In a very real sense, Y(s) is the **solution** to the **initial value problem.**

An initial value problem consists of an nth order ODE together with initial values for the dependent variable and its first n-1 derivatives. Thus, for a third order ODE, we must supply the ODE itself plus values for $y(0+)$, $y'(0+)$, and $y''(0+)$. We take the Laplace transform of the ODE, apply the initial conditions, and obtain Y(s), which is the **Laplace domain** solution. We must **invert** Y(s) to obtain the **time domain** solution:

$$y(t) = \pounds^{-1}[Y(s)] \qquad (4.17)$$

We will shortly study methods for taking the inverse transformation denoted by Equation 4.17. For the moment, we only note the **uniqueness theorem**: for each f(t) there is one and only one F(s), and for each F(s) there is one and only one f(t). Thus, inversion of F(s) provides a unique function f(t).

## 4.3.4 Integrals

Given F(s) = £[f(t)], a simple relationship exists for the Laplace transform of the definite integral:

$$£\left[\int_0^t f(t)dt\right] = \frac{F(s)}{s} \tag{4.18}$$

To show this, it is helpful to use a **dummy variable of integration**, θ:

$$\int_0^t f(t)dt = \int_0^t f(\theta)d\theta$$

The Laplace transform becomes

$$£[\int_0^t f(t)dt] = \int_0^\infty e^{-st} \int_0^t f(\theta)d\theta dt$$

We integrate the outside integral by parts using

$$u = \int_0^t f(\theta)d\theta \qquad du = f(t)dt$$

$$v = -\frac{1}{s}e^{-st} \qquad dv = e^{-st}\,dt$$

Thus,

$$\int_0^\infty e^{-st} \int_0^t f(\theta)d\theta dt = \int_0^\infty udv = \frac{-e^{-st}}{s}\int_0^t e^{-st}\, f(\theta)d\theta \Big|_0^\infty + \frac{1}{s}\int_0^\infty f(t)dt$$

$$= \frac{1}{s}£[f(t)] = \frac{F(s)}{s}$$

Thus, integration in the time domain is equivalent to division by s in the Laplace domain. This result allows solution by Laplace transform techniques of differential-integral equations that sometimes arise in the analysis of engineering systems. It also shows that a Laplace transform of the form $s^{-n}Y(s)$ corresponds to n-fold integration of y(t) in the time domain.

**EXAMPLE 4.6** Obtain the Laplace transform of the following differential-integral equation

$$y'' - by + c\int_0^t y\,dt = 3 \qquad \text{(i)}$$

**Solution:** Applying the equations for derivatives and integrals gives

$$s^2Y(s) - sy(0+) - y'(0+) - bY(s) + \frac{cY(s)}{s} = \frac{3}{s}$$

Solving for Y(s) gives

$$Y(s) = \frac{\dfrac{3}{s} + sy(0+) + y'(0+)}{s^2 - b + c/s} = \frac{s^2y(0+) + sy'(0+) + 3}{s^3 - bs + c} \qquad \text{(ii)}$$

**Second solution:** The integral in Equation i can be eliminated by differentiating with respect to the upper limit. This gives

$$y''' - by' + cy = 0$$

Taking Laplace transforms gives

$$s^3Y(s) - s^2y(0+) - sy'(0+) - y''(0+) - bsY(s) + by(0+) + cY = 0$$

Solution for Y(s) gives

$$Y(s) = \frac{(s^2-b)y(0+) + sy'(0+) - y''(0+)}{s^3 - bs + c} \qquad \text{(iii)}$$

This result is obviously similar to Equation ii but contains the term y''(0+), which is not directly known as an initial condition. To find y''(0+), we evaluate Equation i at t = 0:

$$y''(0+) - by(0+) + 0 = 3$$

Substitution into Equation iii gives Equation ii.

**EXAMPLE 4.7** Find the Laplace transform of

$$f(t) = \int_0^t \int_0^t \sin wt\, dt\, dt \qquad \text{(i)}$$

**Solution:** Table 4.1 gives

$$£[\sin wt] = \frac{\omega}{s^2 + \omega^2}$$

Applying the integral theorem twice gives

$$F(s) = \frac{\omega}{s^2(s^2 + \omega^2)}$$

This is a nice-looking result but it should be verified by explicit integration of Equation i followed by Laplace transformation of the resulting integral. To avoid confusion, different symbols should be used for the dummy variables in an iterated integral. Thus, we write

$$f(t) = \int_0^t \int_0^t \sin \omega\theta\, d\theta d\tau = \frac{1}{\omega} \int_0^t (1 - \cos \omega\tau) d\tau$$

$$= \frac{t}{\omega} - \frac{\sin \omega t}{\omega^2}$$

We now take the Laplace transform using entries in Table 4.1 plus the linearity property.

$$F(s) = £\left[\frac{t}{\omega}\right] - £\left[\frac{\sin \omega t}{\omega^2}\right] = \frac{1}{\omega s^2} - \frac{\omega}{\omega^2(s^2 + \omega^2)}$$

$$= \frac{1}{\omega}\left[\frac{(s^2 + \omega^2) - s^2}{s^2\,(s^2 + \omega^2)}\right] = \frac{\omega}{s^2\,(s^2 + \omega^2)}$$

---

### 4.3.5 Multiplication and Division by t

The Laplace transform is a function of s and can be differentiated as such. Differentiating Equation 4.1 with respect to s gives

$$\frac{dF}{ds} = \int_0^\infty f(t)\,(-t)e^{-st}\,dt$$

or

$$-\frac{dF}{ds} = \int_0^\infty tf(t)e^{-st}\,dt = \pounds[tf(t)] \qquad (4.19)$$

Thus, differentiation in the Laplace domain is equivalent to multiplication by t in the time domain. For an initially relaxed system with $y(0+) = y'(0+) = \ldots = 0$, the results in Section 4.3.1 show that the inverse relation holds: Differentiation in the time domain is equivalent to multiplication by s in the Laplace domain.

Now, instead of differentiating Equation 4.1, suppose we integrate

$$\int_\infty^s F(s)ds = \int_\infty^s \int_0^\infty f(t)e^{-st}\,dt\,ds$$

Reversing the order of integration,

$$\int_\infty^s F(s)ds = \int_0^\infty f(t) \int_\infty^s e^{-st}\,ds\,dt$$

$$= \int_0^\infty f(t)\left(\frac{-1}{t}\right) e^{-st}\,dt$$

or

$$\int_s^\infty F(s)\,ds = \int_0^\infty \frac{f(t)}{t}\,e^{-st}\,dt = \pounds\left[\frac{f(t)}{t}\right] \qquad (4.20)$$

Thus, integration in the Laplace domain is equivalent to division by t in the time domain. The results in Section 4.3.4 show the inverse relation holds: Integration in the time domain is equivalent to division by s in the Laplace domain.

## 4.3.6 Shifts in t and s

We now consider two theorems of Laplace transformation that deal with shifts (offsets) in the transformation variables. We first consider a time domain shift, $t \to t - t_0$ and then consider a Laplace domain shift, $s \to s - s_0$.

The Laplace transform, $\pounds[f(t)]$, can be written as $\pounds[f(t)U(t)]$ without changing its value since $U(t) = 0$ for $t < 0$ and $U(t) = 1$ for $t > 0$. Thus,

$$\pounds[f(t)U(t)] = \int_0^\infty f(t)U(t)e^{-st}dt = \int_0^\infty f(t)e^{-st}dt = F(s) \tag{4.21}$$

We now consider the same integral subject to a shift in t: $\tau = t - t_0$.

$$\pounds[f(t-t_0)U(t-t_0)] = \int_0^\infty f(t-t_0)U(t-t_0)e^{-st}dt$$

$$= \int_0^\infty f(\tau)U(\tau)e^{-s(\tau-t_0)}\,d\tau = e^{-st_0}\int_0^\infty f(\tau)U(\tau)e^{-s\tau}d\tau \tag{4.22}$$

$$= e^{-st_0}\,F(s)$$

Thus, a shift in time corresponds to the factor $e^{-st_0}$ in the Laplace domain. Note that the various functions

$f(t-t_0)$, $f(t-t_0)U(t)$,

$f(t)U(t-t_0)$, $f(t-t_0)U(t-t_0)$

are all different. Equation 4.22 applies only to the last of these functions.

A dual property with Equation 4.22 is the shift in the Laplace transform parameter, $s \to s-s_0$.

$$F(s-s_0) = \int_0^\infty f(t)e^{-(s-s_0)t}\,dt$$

$$= \int_0^\infty e^{s_0t}\,f(t)e^{-st}dt = \pounds[e^{s_0t}\,f(t)] \tag{4.23}$$

## 4.3.7 Scale Changes

Here we again consider dual properties: a scale change in t and a scale change in s. For the change in t,

$$\pounds[f(\alpha t)] = \int_0^\infty f(\alpha t)e^{-st}dt$$

Substitute $\tau = \alpha t$ to obtain

$$\pounds[f(\alpha t)] = \int_0^\infty f(\tau)e^{-\frac{s}{\alpha}\tau}\frac{d\tau}{\alpha} = \frac{F(s/\alpha)}{\alpha} \qquad (4.24)$$

The dual property is

$$F(\beta s) = \int_0^\infty f(t)e^{-\beta st}st$$

Substitute $\tau = \beta t$. Then

$$F(\beta s) = \int_0^\infty f(\tau/\beta)e^{-st}\frac{d\tau}{\beta} = \frac{1}{\beta}\pounds[f(\tau/\beta)]$$

Since $\tau$ is a dummy variable of integration we can also write this result as

$$F(\beta s) = \pounds[\frac{1}{\beta}f(t/\beta)] \qquad (4.25)$$

## 4.3.8 Initial and Final Values

Here is another beautiful duality between functions in the time and Laplace domains. An initial value in the time domain ($t \to 0$) is a final value ($s \to \infty$) in the Laplace domain. Conversely, an initial value in the Laplace domain ($s \to 0$) is a final value ($t \to \infty$) in the time domain. Specifically,

$$\lim_{t \to 0+} f(t) = f(0+) = \lim_{s \to \infty} sF(s) \qquad (4.26)$$

and

$$\lim_{t \to \infty} f(t) = f(\infty) = \lim_{s \to 0} sF(s) \qquad (4.27)$$

We prove only Equation 4.27. Begin by supposing both f(t) and its first derivative are Laplace transformable. Take the limit as $s \to 0$ of Equation 4.13 to obtain

$$\lim_{s \to 0} \int_0^\infty f' e^{-st} dt = \lim_{s \to 0} [sF(s) - f(0+)]$$

The limit of the integral is

$$\lim_{s \to 0} \int_0^\infty f' e^{-st} dt = \int_0^\infty f'(t) dt = f(t) \Big|_0^\infty = f(\infty) - f(0+)$$

The f(0+) terms cancel between these results to give Equation 4.27.

The final value theorem may not apply to situations where f(t) is periodic. An example where it does not apply is f(t) = sin ωt. It may also fail when f(∞) does not exist. That is, when f(∞) = ± ∞. See Example 5.2.

### 4.3.9 Convolution

This is a final and fantastic property of Laplace transformation. We consider two initially relaxed functions, $f_1(t)$ and $f_2(t)$. Their **convolution** is defined as

$$f_1 * f_2 = \int_0^t f_1(t-\tau) f_2(\tau) dt = \int_0^t f_1(\tau) f_2(t-\tau) d\tau \tag{4.28}$$

The equivalence of these two integrals can be demonstrated by a simple change of variables: τ = t - τ. The **convolution**, $f_1 * f_2$, can be Laplace transformed.

$$£[f_1 * f_2] = £\left[\int_0^t f_1(t-\tau)\, f_2(\tau) d\tau\right] = £\left[\int_0^t f_1(\tau) f_2(t-\tau) d\tau\right] = F_1(s)\, F_2(s)$$

The final result,

$$£[f_1 * f_2] = F_1\, F_2 \tag{4.29}$$

is that the Laplace transform operator transforms convolution into multiplication. The proof of 4.29 is fairly straightforward but lengthy.

## 4.4 INVERSION

The previous sections of this chapter have shown how to take the Laplace transform of various time-domain functions. This methodology allows us to translate an ordinary differential equation with t as the independent variable into an algebraic equation with s as the independent variable. Initial conditions are applied as part of the transformation, and the final result, Y (s) = £[y(t)], is the complete solution to the governing ODE. By complete solution we now mean that the characteristic (or complementary) solution, the particular solution, and all initial conditions are included. What remains is to find the **inverse transformation**, $y(t) = \pounds^{-1}[Y(s)]$, so that the solution is obtained in the time domain.

The inversion problem has been solved in a very general manner:

$$f(t) = \frac{1}{2\pi i} \int_{\sigma-i\omega}^{\sigma+i\omega} F(s)e^{st}ds \tag{4.30}$$

Equations 4.1 and 4.30 constitute a **Laplace transform pair**. Given f(t) we use Equation 4.1 to find F(s), and given F(s) we use Equation 4.30 to find f(t).

The integration tools needed to evaluate Equation 4.1 are the ordinary kind and, explicitly at least, involve only real-valued functions. The integration in Equation 4.40 takes place in the complex plane and requires the theory of residues. This theory is given in mathematics courses on functions of a complex variable. The present book uses simpler approaches to inversion. These simpler approaches are consistent with Equation 4.30, but do not require a complete understanding of Equation 4.30 in order to be used.

The last statement raises an interesting point in the philosophy of engineering, science, and applied mathematics. To what extent must a worker know his tools? The use of poorly understood methodologies can lead to wrong answers with sometimes serious consequences. As a general rule, one must know **how** a tool works to ensure that it **does** work; but there is an exception to this general rule. We can ensure that a tool does work by always verifying its result. This is possible in the case of analytical solutions to ordinary differential equations. Suppose some formula $y = y(t)$ is claimed to be the solution to an initial value problem. This claim can be readily varied by checking. Do the initial values $y(0+)$, $y'(0+)$, ... agree with those in the problem statement? Does $y(t)$ satisfy the governing ODE? If so, then $y(t)$ is indeed the solution and it remains the solution regardless of how it was found. In this sense, even intuition is an acceptable means of solving a problem provided the solution can be verified.

Much of **operational calculus**, which includes Laplace transform techniques, was developed by engineers and scientists using fairly intuitive methods. The most useful inversion methods were invented by an electrical engineer named Heaviside. His methods worked even though his mathematical arguments were not quite rigorous.

As we have seen, numerical solutions to ODEs are almost always easier to obtain than analytical solutions. However, they are much harder to verify. We have stressed a simple, marching ahead technique because its working principle is based on the definition of a derivative and because the computer program is extremely easy for this method. However, it (and the more complicated methods) can go wrong in special cases. Good numerical solutions may require overly small $\Delta t$ or overly high precision for a given computer. The best verification is that the numerical solution agrees with analytical solutions, at least in those special cases where an analytical solution is possible.

### 4.4.1 Transform Tables

The simplest way to invert a Laplace transform is to find F(s) in a table of transforms such as Table 4.1. Then the corresponding f(t) is just read from the table, perhaps with a little algebra to find the coefficients.

**EXAMPLE 4.8** Find the inverse transform of

$$F(s) = \frac{s+5}{s^2 - 2s - 2}$$

**Solution:** The first step is to factor the denominator:

$$s^2 - 2s - 2 = (s - 1 - \sqrt{3})(s - 1 + \sqrt{3})$$

and

$$F(s) = \frac{s + 5}{(s-1-\sqrt{3})(s-1+\sqrt{3})} \qquad \text{(i)}$$

so that there are two distinct, real roots. These roots are called **poles** of F(s). Table 4.1 gives the transform pair

$$f(t) = Ae^{-\alpha t} + Be^{-\beta t}, \quad F(s) = \frac{[A+B]s + [A\beta + B\alpha]}{(s + \alpha)(s + B)}$$

This form for F(s) fits Equation i provided

$$\alpha = -1 - \sqrt{3}$$

$$\beta = -1 + \sqrt{3}$$

$$A + B = 1$$

$$A\beta + B\alpha = 5$$

Solving for A and B gives $A = 1/2 + \sqrt{3}$, $B = 1/2 - \sqrt{3}$. Thus,

$$f(t) = (\frac{1}{2} + \sqrt{3})\, e^{(1+\sqrt{3})t} + (\frac{1}{2} - \sqrt{3})\, e^{(1-\sqrt{3})t}$$

is the desired inverse.

---

**EXAMPLE 4.9** Find the inverse of

$$F(s) = \frac{-2s}{s^2 + 2s + 2} \qquad \text{(i)}$$

**Solution:** Factoring the denominator gives complex roots in a conjugate pair:

$$F(s) = \frac{-2s}{(s+1+i)\,(s+1-i)} \qquad \text{(ii)}$$

Table 4.1 contains the transform pair

$$f(t) = e^{-\sigma t}\,[A\sin \omega t + B \cos \omega t], \quad F(s) = \frac{Bs + (A\omega + B\sigma)}{(s + \sigma + \omega i)\,(s + \sigma - \omega i)}$$

Comparison with Equation ii gives $\sigma = \omega = 1$, $B = -2$, $A = +2$. Thus,

$$f(t) = 2e^{-t}\,[\sin t - \cos t]$$

---

**EXAMPLE 4.10** Invert

$$F(s) = \frac{3s}{(s + 1)\,(s - 1\ )}$$

**Solution:** This fits the form

$$f(t) = Ae^{-\alpha t} + Be^{-\beta t}, \quad F(s) = \frac{[A + B]s + [A\beta + B\alpha]}{(s + \alpha)\,(s + \beta)}$$

with $\alpha = 1$, $\beta = -1$, $A + B = 3$, $-A + B = 0$. Thus,

$$f(t) = 1.5e^{-t} + 1.5e^{+t}$$

It happens that

$$\cosh \alpha t = \frac{e^{\alpha t} + e^{-\alpha t}}{2}$$

so that an equivalent solution is

$$f(t) = 3 \cosh t$$

Most people would consider this answer more elegant and more insightful than our first solution. One advantage of using (large) tables of Laplace transforms is that they directly tabulate transform pairs such as

$$f(t) = \cosh \alpha t, \quad F(s) = \frac{s}{s^2 - \alpha^2}$$

and

$$f(t) = \sin \alpha t \sinh \alpha t, \quad F(s) = \frac{2a^2 s}{s^4 + 4a^4}$$

---

## 4.4.2 Heaviside's Expansion Theorem

We now explain a general method for inverting Laplace transforms of the form

$$F(s) = \frac{N(s)}{D(s)} = \frac{\text{Polynomial in degree m in s}}{\text{Polynomial in degree n is s}} \tag{4.31}$$

where the indicated polynomials have real coefficients and where m and n are non-negative integers: $m > 0, n > 0$. There are three cases to be considered:

(i) $m < n$. These are called **proper polynomial fractions.**

(ii) $m > n$. These are **improper polynomial fractions.**

(iii) $m = n$. Some authors call these improper; others call them proper. Let us call them **fractions of equal degree polynomials**, which is long but unambiguous.

Our theory deals primarily with proper fractions, $m < n$, and almost exclusivity with fractions which are not improper, $m \leq n$.

The denominator in Equation 4.31 can be factored to give

$$F(s) = \frac{N(s)}{D(s)} = \frac{N(s)}{(s-s_1)(s-s_2) \ldots (s-s_n)} \tag{4.32}$$

where the $s_j$ are the roots, real or imaginary, of $D(s) = 0$. This factorization supposes the leading coefficient of D(s) is 1. This can always be achieved by dividing each term in N(s) and D(s) by the leading coefficients of D(s). We now suppose F(s) is a strictly proper fraction, $m < n$, and consider three cases:

Case I: The roots of D(s) = 0 are real and distinct.

Case II: The roots of D(s) = 0 contain some repetition, $s_j = s_k$, for some j and k.

Case III: The roots of D(s) = 0 include some imaginary roots in conjugate pairs.

**Case I.** The quotient, N(s)/D(s), can be expanded as the sum of partial fractions:

$$F(s) = \frac{N(s)}{D(s)} = \frac{K_1}{s-s_1} + \frac{K_2}{s-s_2} + \cdots + \frac{K_n}{s-s_n} \tag{4.33}$$

where $K_1$, $K_2$, ..., $K_n$ are constants to be determined. The fact that a proper polynomial fraction can indeed be expanded according to Equation 4.33 is a theorem in algebra which we accept without proof. There are two different approaches to determining the K. One approach is to multiply each term in the expansion by $s-s_j$ where j represents some specific root of D(s) = 0. We then set $s = s_j$ so that the term $s-s_j$ vanishes. Then the various terms in Equation 4.33 vanish except for the jth term. Thus,

$$K_j = \lim_{s \to s_j} \frac{(s-s_j)\, N(s)}{D(s)} \tag{4.34}$$

The limit in Equation 4.34 can be taken in two different ways. For the first way, note that $s - s_j$ is a (nonrepeated) factor of D(s). Thus, canceling the $(s-s_j)$ terms gives

$$K_j = \frac{N(s_j)}{\prod_{i \neq j} (s_j - s_i)} \tag{4.35}$$

The second approach to determining $K_j$ rests on a property of L'Hospital's rule: the rule works whether or not you need to use it. In Equation 4.34, the factor $(s-s_j)$ appears in both numerator and denominator. We canceled them to obtain Equation 4.35. We now retain them so that the limit in Equation 4.34 has the indeterminate form 0/0. Applying L'Hospitals rule gives

$$K_j = \lim_{s \to s_j} \frac{N(s) + (s-s_j)\, N'(s)}{D'(s)} = \lim_{s \to s_j} \frac{N(s_j)}{D'(s_j)} \tag{4.36}$$

The marvelous result can remain true even when D(s) is a power series of infinite degree. When D(s) is a polynomial of degree n > m, Equation 4.36 is equivalent to Equation 4.35. You can use whichever is more convenient.

Once $K_j$ has been evaluated according to Equations 4.35 or 4.36, the sum of partial fractions, Equation 4.33, can be inverted term by term to give

$$f(t) = \pounds^{-1}[F(s)] = \sum_{j=1}^{n} K_j e^{s_j t} \tag{4.37}$$

**EXAMPLE 4.11** Invert

$$F(s) = \frac{N(s)}{D(s)} = \frac{s+4}{2s^2+5s+3}$$

using both forms of Heaviside's expansion theorem.

**Solution:**

$$F(s) = \frac{.5s + 2}{s^2 + 2.5s + 1.5} = \frac{.5s + 2}{(s+1)(s+1.5)}$$

where we have divided both numerator and denominator by the leading coefficient of D(N). We also have factored the denominator to give roots $s_1 = -1$ and $s_2 = -1.5$. The partial sum expansion is

$$F(s) = \frac{K_1}{s+1} + \frac{K_2}{s+1.5}$$

where $K_1$ and $K_2$ can be determined by either of Equations 4.35 or 4.36. Using Equation 4.35 gives

$$K_1 = \frac{.5s_1 + 2}{s_1 - s_2} = \frac{-.5+2}{-1+1.5} = +3$$

$$K_2 = \frac{.5s_2 + 2}{s_2 - s_1} = \frac{-.75+2}{-1.5+1} = \frac{1.25}{-.5} = -2.5$$

Equation 4.37 gives

$$f(t) = 3e^{-t} - 2.5e^{-1.5t}$$

The alternative form is found by differentiating D(s).

$$D' = \frac{d}{ds}(2s^2 + 5s + 3) = 4s + 5$$

Using Equation 4.36

$$K_1 = \frac{s_1 + 4}{4s_1 + 5} = \frac{3}{1} = 3$$

$$K_2 = \frac{s_2 + 4}{4s_2 + s} = \frac{2.5}{-1} = -2.5$$

which is the same result as obtained with Equation 4.35.

---

**Case II.** We now suppose that one of the roots of D(s) = 0 is repeated. Specifically, we suppose root #1 has multiplicity p so that

$$D(s) = (s-s_1)^p\,(s-s_2)\,\ldots\,(s-s_n) \tag{4.38}$$

where we have supposed that D(s) has been normalized by dividing by its leading coefficient.

The partial sum we now seek is

$$F(s) = \frac{K_{11}}{s - s_1} + \frac{K_{12}}{(s-s_1)^2} + \cdots + \frac{K_{1p}}{(s-s_1)^p} + \frac{K_2}{s-s_2} + \cdots + \frac{K_n}{s-s_n} \tag{4.39}$$

For the unrepeated roots, $j > 1$, Equations 4.35 and 4.36 hold as before. For $j = 1$, there are p different values for partial sum constants, $K_{11}$, $K_{12}$, ..., $K_{1p}$. These can be found using the following formula:

$$K_{1k} = \frac{1}{(p-k)!}\,\frac{d^{(p-k)}}{ds^{(p-k)}}\left[(s-s_1)^p\,\frac{N(s)}{D(s)}\right]_{s_1},\quad k = 1, 2, \ldots, p \tag{4.40}$$

---

**EXAMPLE 4.12** Invert

$$F(s) = \frac{1}{(s+1)^3}$$

**Solution:** There is a single root with multiplicity p = 3. Equation 4.40 gives

$$K_{1k} = \frac{1}{(3-k)!}\frac{d^{3-k}}{ds^{3-k}}\left[\frac{(s+1)^3\,(1)}{(s+1)^3}\right] = \frac{1}{(3-k)!}\frac{d^{3-k}(1)}{ds^{3-k}}$$

Only the k = 3 terms survive differentiation. The partial sum expansion has $K_{11} = K_{12} = 0$ and $K_{13} = 1$ so that this is a trivial case where the partial sum expansion is the same as the original fraction.

The inverse transform uses the following transform pair from Table 4.1:

$$f(t) = t^n e^{-\alpha t} \qquad F(s) = \frac{n!}{(s+\alpha)^{n+1}}$$

To invert, we set n = 2 and transpose the factor of n! Thus,

$$f(t) = \frac{t^2 e^{-\alpha t}}{2!}$$

---

**EXAMPLE 4.13** Find the inverse of

$$F(s) = \frac{1}{(s+1)^2\,(s-2)}$$

**Solution:** Here the roots are $s_1 = s_2 = -1$, repeated with p = 2, and $s_3 = +2$. Equation 4.40 gives

$$K_{1k} = \frac{1}{(2-k)!} \frac{d^{2-k}}{dt^{2-k}} \left[ \frac{(s+1)^2}{(s+1)^2 (s-2)} \right]_{s=-1}$$

$$= \frac{1}{(2-k)!} \frac{d^{2-k}}{dt^{2-k}} \left[ \frac{1}{s-2} \right]_{s=-1}$$

$$= \frac{1}{1!} \left( \frac{1}{s-2} \right)_{s=-1} = -\frac{1}{3} \text{ for } k = 2$$

$$= \frac{1}{0!} \left( \frac{-1}{(s-2)^2} \right)_{s=-1} = \frac{-1}{9} \text{ for } k = 1$$

The partial sum expansion is

$$\frac{-1}{9(s+1)} + \frac{-1}{3(s+1)^2} + \frac{K_3}{(s-2)}$$

We use Equation 4.44 to find

$$K_3 = \frac{N(s_3)}{\prod\limits_{i \neq j} (s_j - s_i)} = \frac{1}{(s_3 - s_1)^2} = \frac{1}{9}$$

Thus, the partial sum expansion is

$$F(s) = \frac{-1}{9(s+1)} + \frac{-1}{3(s+1)^2} + \frac{1}{9(s-2)}$$

and the inverse is

$$f(t) = \frac{-e^{-t}}{9} - \frac{e^{-t}}{3} + \frac{e^{2t}}{9}$$

---

**Case III.** In this case, two of the roots are complex conjugates so that D(s) must be at least second order. If D(s) is exactly second order,

$$\begin{aligned} D(s) &= (s + \sigma + \omega i)(s + \sigma - \omega i) = (s + \sigma)^2 + \omega^2 \\ &= s^2 + 2\sigma s + \sigma^2 + \omega^2 \end{aligned} \tag{4.41}$$

The transform pair corresponding to this case is

$$f(t) = e^{-\sigma t}[A\sin\omega t + B\cos\omega t],\ F(s) = \frac{Bs + (A\omega + B\sigma)}{(s+\sigma)^2 + \omega^2} \tag{4.42}$$

When D(s) is factored, $\sigma$ and $\omega$ can be found by inspection. Then, since we have assumed $n = 2$ and are considering only the case of $m < n$, N(s) must be linear in s

$$N(s) = N_1 s + N_2 \tag{4.43}$$

Comparing this with Equation 4.42 gives

$$B = N_1, \qquad A = (N_2 - N_1\sigma)/\omega \tag{4.44}$$

**EXAMPLE 4.14** Find the inverse of

$$F(s) = \frac{s}{s^2 + 2s + 5}$$

**Solution:** The roots of D(s) = 0 can be found by the quadratic formula. The result is

$$F(s) = \frac{s}{(s + 1 + 2i)\,(s + 1 - 2i)}$$

so that $\sigma = 1$, $\omega = 2$, $N_1 = 1$, and $N_2 = 0$. Equation 4.44 gives B = 1 and A = -1/2. Thus, the solution is

$$f(t) = e^{-t}[-\frac{1}{2}\sin 2t + \cos 2t]$$

---

The reader will note that Equation 4.44 was derived using tabulated transforms, Table 4.1, rather than Heaviside's expansion theorem. The expansion theorem gives the same result but requires a modest amount of complex arithmetic.

**EXAMPLE 4.15** Use Heaviside's expansion theory to invert

$$F(s) = \frac{s}{(s + 1 + 2i)\,(s + 1 - 2i)}$$

**Solution:** D(s) is already factored and the roots are distinct. Equation 4.44 gives

$$K_1 = \frac{s_1}{s_1 - s_2} = \frac{-1-2i}{-1-2i+1-2i} = \frac{1+2i}{4i} = \frac{1}{2} - \frac{i}{4}$$

$$K_2 = \frac{s_2}{s_2 - s_1} = \frac{-1+2i}{-1+2i+1+2i} = \frac{-1+2i}{4i} = \frac{1}{2} + \frac{i}{4}$$

Note that $K_1$ and $K_2$ form a conjugate pair. This follows from the fact that $s_1$ and $s_2$ are a conjugate pair. The partial sum expansion is

$$F(s) = \frac{1/2 - i/4}{s+1-2i} + \frac{1/2 + i/4}{s+1+2i}$$

The inverse is

$$f(t) = \left(\frac{1}{2} + \frac{i}{4}\right)e^{(-1+2i)t} + \left(\frac{1}{2} - \frac{i}{4}\right)e^{(-1-2i)t}$$

$$= e^{-t}\left[\left(\frac{1}{2} + \frac{i}{4}\right)e^{2it} + \left(\frac{1}{2} - \frac{i}{4}\right)e^{-2it}\right]$$

To proceed from this point, two fundamental identities are needed:

$$\sin \omega t = \frac{1}{2i}\left[e^{+i\omega t} - e^{-i\omega t}\right]$$

$$\cos \omega t = \frac{1}{2}\left[e^{+i\omega t} + e^{-i\omega t}\right]$$

so that

$$-\frac{1}{2}\sin \omega t + \cos \omega t = \frac{-e^{+i\omega t}}{4i} + \frac{e^{-i\omega t}}{4i} + \frac{e^{+i\omega t}}{2} + \frac{e^{-i\omega t}}{2}$$

$$= \left(\frac{1}{2} + \frac{i}{4}\right) e^{+i\omega t} + \left(\frac{1}{2} - \frac{i}{4}\right) e^{-i\omega t}$$

We thus set $\omega = 2$ to obtain

$$f(t) = e^{-t}\left[-\frac{1}{2}\sin 2t + \cos 2t\right]$$

in agreement with the previous example.

---

The reader will appreciate the relative ease with which Example 4.14 was worked. For roots in conjugate pairs, it is easier to invert using the cookbook recipes of Equations 4.41 – 4.44 than by using Heaviside's expansion theorem. When the expansion theorem is used, the quadratic factor containing the conjugate pair is retained as a single term in the expansion. Suppose roots 1 and 2 form a conjugate pair and that higher roots are distinct. Then

$$F(s) = \frac{N(s)}{D(s)} = \frac{k_1 + k_2 s}{(s + \sigma + \omega i)(s + \sigma - \omega i)} + \frac{k_3}{s - s_3} + \cdots$$

$$= \frac{k_1 + k_2 s}{s^2 + 2\sigma s + \sigma^2 + \omega^2} + \frac{k_3}{s - s_3} \qquad (4.45)$$

Expansion in this form avoids the use of complex arithmetic.

**EXAMPLE 4.16** Invert

$$F(s) = \frac{1}{3s^2 (s^2 + 4)} = \frac{1}{3s^2 (s + 2i)(s - 2i)} \qquad (i)$$

**Solution:** This is not exactly the situation of Equation 4.45 since $s = 0$ has multiplicity $p = 2$. However, we merely combine the methodology of Cases II and III. The partial sum is

$$F(s) = \frac{1}{3}\left[\frac{K_{10}+K_{11}s}{s^2 + 4} + \frac{K_{21}}{s} + \frac{K_{22}}{s^2}\right] \qquad (ii)$$

There are four coefficients to be determined. We could use Equations 4.35, 4.36, or 4.40 as in previous examples. However, a direct algebraic approach is also possible. We collect terms in Equation ii using the common denominator:

$$F(s) = \frac{1}{3}\left[\frac{(K_{10} + K_{11}s)\, s^2 + K_{21}s\,(s^2 + 4) + K_{22}\,(s^2 + 4)}{s^2\,(s^2 + 4)}\right]$$

$$= \frac{1}{3}\left[\frac{(K_{11} + K_{21})s^3 + (K_{10} + K_{22})s^2 + 4K_{21}s + 4K_{22}}{s^2(s^2 + 4)}\right] \qquad (iii)$$

Equating powers of s between Equations i and iii gives

$$K_{11} + K_{12} = 0$$

$$K_{10} + K_{22} = 0$$

$$4K_{21} = 0$$

$$4K_{22} = 1$$

This set has the solutions $K_{10}$= -1/4, $K_{11}$ = 0, $K_{21}$ = 0, and $K_{22}$ = 1/4. Thus, the partial fraction expansion is

$$F(s) = \frac{1}{3}\left[\frac{-1/4}{s^2+4} + \frac{1/4}{s^2}\right] = \frac{1}{12}\left[\frac{-1}{s^2+4} + \frac{1}{s^2}\right]$$

We invert the first term using Equation 4.42 with $\sigma = 0$, $\omega = 2$, A = -1/2, and B = 0. The second term, $s^{-2}$, inverts to t. Thus,

$$f(t) = \frac{1}{12}\left[\frac{\sin 2t}{-2} + t\right]$$

---

### 4.4.3 Fractions of Equal Degree Polynomials

Returning to Equation 4.31, it sometimes happens that N(s) and D(s) have equal degrees, m = n. In this case, F(s) is expanded as

$$F(s) = K_0 + \frac{K_1}{s + s_1} + \frac{K_2}{s + s_2} + \cdots \tag{4.46}$$

where $s_1$, $s_2$, ... are the roots of D(s) = 0. Values for $K_0$, $K_1$, ... can be found by writing simultaneous equations for the coefficients of $s^j$ as in Example 4.16 or by using Equations 4.35, 4.36, or 4.40. Equation 4.46 is written as if all the roots were distinct, but it can be modified easily to handle cases with repeated (or conjugate) roots. What is important about Equation 4.46 is the first term, $K_0$. This term exists if m = n. The inverse of Equation 4.46 is

$$f(t) = K_0\delta(t) + K_1e^{-s_1t} + \ldots \tag{4.47}$$

**EXAMPLE 4.17** Invert

$$F(s) = \frac{2s^2 + 7s + 8}{s^2 + 3s + 2} = \frac{2s^2 + 7s + 8}{(s+1)(s+2)}$$

**Solution:**

$$\frac{2s^2 + 7s + 8}{(s+1)(s+2)} = K_0 + \frac{K_1}{s+1} + \frac{K_2}{s+2}$$

$$= \frac{K_0(s^2 + 3s + 2) + K_1(s+2) + K_2(s+1)}{(s+1)(s+2)}$$

Equating coefficients of $s^j$ gives

$$K_0 = 2$$

$$3K_0 + K_1 + K_2 = 7$$

$$2K_0 + 2K_1 + K_2 = 8$$

The solution is $K_0 = 2$, $K_1 = 3$, and $K_2 = -2$. Thus,

$$F(s) = 2 + \frac{3}{s+1} - \frac{2}{s+2}$$

The inverse is

$$f(t) = 2\delta(t) + 3e^{-t} - 2e^{-2t}$$

---

The partial sum approach can be extended to the case of m > n. Additional terms like $K_{-1}s$, $K_{-2}s^2$, ... would have to be included in the sum. The various coefficients could be evaluated by the algebraic route demonstrated above. However, the inverse transforms for terms like $K_{-1}s$ and $K_{-n}s^n$ are rarely listed in tables of transforms. From a strictly formal viewpoint, the inverse of $s^n$ corresponds to the nth derivative of the delta function, $\delta(t)$. Such concepts are possible but have few engineering applications.

### 4.4.4 Recap of Partial Fraction Expansions

To invert F(s) = N(s)/D(s), do the following:

1. Factor D(s) into its various linear factors: $(s+\alpha)\ (s+\beta)\ (s+\gamma) \dots$ . However, do not factor those quadratic terms which give imaginary or complex roots. A fairly general result after factoring is

$$D(s) = (s+\alpha)\,(s+\beta)^p\,(s^2 + 2s\sigma + \sigma^2 + \omega^2)$$

where the $(s+\alpha)$ term corresponds to a single real root at $s = -\alpha$, the $(s+\beta)^p$ term corresponds to repeated real roots at $s = -\beta$, and the $(s^2 + 2s\sigma + \sigma^2 + \omega^2)$ term corresponds to complex conjugate roots at $s = -\sigma \pm i\omega$.

2. Write a partial fraction expansion for N(s)/D(s) using undetermined coefficients.

2.1 For each unrepeated real root, there is a single term in the expansion:

$$\frac{K_1}{s + \alpha}$$

2.2 For repeated real roots, there are p undetermined coefficients in the expansion:

$$\frac{K_{21}}{(s+\beta)} + \frac{K_{22}}{(s+\beta)^2} + \dots + \frac{K_{2p}}{(s+\beta)^p}$$

2.3 For a conjugate pair, the expansion contains two coefficients

$$\frac{K_{30} + K_{31}s}{(s^2 + 2s\sigma + \sigma^2 + \omega^2)}$$

2.4 At this point, the expansion will contain n undetermined coefficients where n is the degree of D(s). If it happens that N(s) is also of degree n, add a $K_0$ term to the expansion.

3. Evaluate the "easy" coefficients in the partial fraction expansion.

3.1 For each single root, $s_1 = -\alpha$, multiply by s+a and then set $s = -\alpha$. This allows $K_1$ to be found from Equation 4.35.

3.2 For each repeated real root, $s_2 = -\beta$, multiple by $(s+\beta)^p$ and then set $s = -\beta$. This allows $K_{2p}$ to be found from Equations 4.35.

4. At this point, there are usually only a few coefficients yet to be determined. The best general method is to collect the undetermined coefficients over the common denominator, D(s), and then to equate coefficients on the various powers of s. This gives a set of linear algebraic equations for the unknown coefficients.

   For highly repeated roots, it may be advantageous to use the differentiation formula, Equation 4.36.

5. The various terms in the partial fraction are inverted term by term using Table 4.1.

A truly general example leads to very tedious algebra. In fact, it would provide good motivation for learning one of the computer programs that do algebraic manipulations. The following example illustrates the general scheme without getting bogged down in the algebra.

**EXAMPLE 4.18** Invert

$$\frac{N(s)}{D(s)} = \frac{s^3 - 3}{(s+1)\,(s-1)^2}$$

**Solution:** The expansion uses Steps 2.1, 2.2, and 2.4 to give

$$\frac{s^3-3}{(s+1)\,(s-1)^2} = K_0 + \frac{K_1}{s+1} + \frac{K_{21}}{s-1} + \frac{K_{22}}{(s-1)^2}$$

Step 3.1 is used to find $K_1$. Specifically, multiply by s+1 and set $s_1 = -1$ to obtain

$$K_1 = \frac{(-1)^3 - 3}{(-1-1)^2} = -1$$

Step 3.2 is used to find $K_{22}$. Specifically, multiply by $(s-1)^2$ and set $s_2 = +1$ to obtain

$$K_{22} = \frac{(+1)^3 - 3}{(+1+1)} = -1$$

Step 4 is now applied. The partial fraction is

$$\frac{s^3 - 3}{(s+1)(s-1)^2} = K_0 + \frac{-1}{(s+1)} + \frac{K_{21}}{s-1} + \frac{-1}{(s-1)^2}$$

Collecting over D(s) as the common denominator gives

$$\frac{s^3-3}{(s+1)(s-1)^2} = \frac{K_0\,(s+1)\,(s-1)^2 - (s-1)^2 + K_{21}(s+1)\,(s-1) - (s+1)}{(s+1)\,(s-1)^2}$$

or, after some algebra,

$$s^3 + s^2 - s - 1 = K_0 s^3 + (-K_0 + K_{21})s^2 - K_0 s + (K_0 - K_{21})$$

Equating coefficients on powers of s gives

$$\begin{aligned} 1 &= K_0 && (s^3 \text{ term}) \\ 1 &= -K_0 + K_{21} && (s^2 \text{ term}) \\ -1 &= -K_0 && (s \text{ term}) \\ -1 &= K_0 - K_{21} && (s^0 \text{ term}) \end{aligned}$$

Solving the first two members of this set gives $K_0 = 1$, $K_{21} = 2$. (The second two members are also satisfied. This confirms our results for $K_1$ and $K_{22}$.) At this point we have

$$\frac{N(s)}{D(s)} = 1 - \frac{1}{s+1} + \frac{2}{s-1} - \frac{1}{(s-1)^2}$$

Term-by-term inversion gives

$$\mathcal{L}^{-1}\left[\frac{N(s)}{D(s)}\right] = \delta(t) - e^{-t} + 2e^{t} - te^{t}$$

---

**TABLE 4.1**
**FUNCTION-TRANSFORM PAIRS**

| $f(t),\ t \geq 0$ | $F(s) = \pounds[f(t)]$ |
|---|---|
| $\delta(t)$ | $1$ |
| $U(t)$ | $\frac{1}{s}$ |
| $t^n\ (n = 1,2,\cdots)$ | $n!/s^{n+1}$ |
| $e^{-\alpha t}$ | $\frac{1}{s+\alpha}$ |
| $te^{-\alpha t}$ | $\frac{1}{(s+\alpha)^2}$ |
| $t^n e^{-\alpha t}$ | $\frac{n!}{(s+\alpha)^{n+1}}$ |
| $Ae^{-\alpha t} + Be^{-\beta t}$ | $\frac{[A+B]s + [A\beta+B\alpha]}{(s+\alpha)(s+\beta)}$ |
| $(A + Bt)e^{-\alpha t}$ | $\frac{As + [A\alpha + B]}{(s+\alpha)^2}$ |
| $\sin \omega t$ | $\frac{\omega}{s^2+\omega^2} = \frac{\omega}{(s+\omega i)(s-\omega i)}$ |
| $\cos \omega t$ | $\frac{s}{s^2+\omega^2} = \frac{s}{(s+\omega i)(s-\omega i)}$ |
| $A \sin \omega t + B \cos \omega t$ | $\frac{Bs + A\omega}{s^2+\omega^2}$ |
| $e^{-\sigma t} \sin \omega t$ | $\frac{\omega}{(s+\sigma)^2+\omega^2} = \frac{\omega}{(s+\sigma+i\omega)(s+\sigma-i\omega)}$ |
| $e^{-\sigma t} \cos \omega t$ | $\frac{s+\sigma}{(s+\sigma)^2+\omega^2} = \frac{s+\sigma}{(s+\sigma+i\omega)(s+\sigma-i\omega)}$ |
| $e^{-\sigma t}[A \sin \omega t + B \cos \omega t]$ | $\frac{Bs + A\omega + B\sigma}{(s+\sigma)^2+\omega^2} = \frac{Bs + (A\omega+B\sigma)}{(s+\sigma+i\omega)(s+\sigma-i\omega)}$ |
| $\sin \omega t - \omega t \cos \omega t$ | $\frac{2\omega^2}{(s^2+\omega^2)^2}$ |
| $t \sin \omega t$ | $\frac{2\omega s}{(s^2+\omega^2)^2}$ |
| $t \cos \omega t$ | $\frac{s^2-\omega^2}{(s^2+\omega^2)^2}$ |

**TABLE 4.2**
**OPERATION - TRANSFORM PAIRS**

| Operation | f(t) | F(s) = £[F(t)] |
|---|---|---|
| First Derivative | $f' = \frac{df}{dt}$ | $sF(s) - f(0+)$ |
| Second Derivative | $f'' = \frac{d^2f}{dt^2}$ | $s^2F(s) - sf(0+) - f'(0+)$ |
| nth Derivative | $f^{(n)} = \frac{d^nf}{dt^n}$ | $s^nF(s) - s^{n-1} f(0+) - \cdots - f^{(n-1)}(0+)$ |
| Multiplication by t | $tf(t)$ | $-\frac{dF}{ds} = -F'(s)$ |
| Division by t | $\frac{f(t)}{t}$ | $\int_0^\infty F(s)ds$ |
| Integration | $\int_0^t f(t)dt$ | $\frac{F(s)}{s}$ |
| Shift in t | $f(t-t_0)U(t-t_0)$ | $e^{-t_0 s} F(s)$ |
| Shift in s | $e^{s_0 t} f(t)$ | $F(s-s_0)$ |
| Scale Change in t | $f(\alpha t)$ | $\frac{F(s/\alpha)}{\alpha}$ |
| Scale Change in s | $\frac{f(t/\beta)}{\beta}$ | $F(\beta s)$ |
| Initial Value | $\lim_{t\to 0} f(t) = f(0+)$ | $\lim_{s\to\infty} sF(s)$ |
| Final Value | $\lim_{t\to\infty} f(t) = f(\infty)$ | $\lim_{s\to 0} sF(s)$ |
| Convolution | $f_1*f_2 = \int_0^t f_1(\tau)f_2(t-\tau)d\tau = \int_0^t f_1(t-\tau)f_2(\tau)d\tau$ | $F_1(s)F_2(s)$ |

## PROBLEMS

4.1 Determine the Laplace transform of $f(t) = \sinh \alpha t$.

4.2 Find the inverse of

$$F(s) = \frac{1}{s^4 - \alpha^4}$$

4.3 Invert

$$F(s) = \frac{s+1}{s+2}$$

4.4 Evaluate the convolution integral

$$I = \int_0^t e^{-\alpha(t-\tau)} e^{-\beta\tau} \, d\tau$$

a) By direct integration

b) Using Laplace transforms

4.5 Invert

$$F(s) = \frac{1}{s^3 + 3s^2 + s - 5}$$

4.6 Find $f(0+)$ and $f(\infty)$ given

$$F(s) = \frac{4s^2 + 1}{s(s+1)(s+2)}$$

4.7 Find the Laplace transform of

$$a'' - 1 = e^{-t}$$

given that $a = a' = 0$ at $t = 0$. Invert your result to find $a(t)$.

4.8 Invert

$$\frac{e^{-3s}}{s^2}$$

4.9 For some simple function $f(t)$ and for $t_0 > 0$, draw plots of $f(t-t_0)$, $f(t-t_0)U(t)$, $f(t)U(t-t_0)$, and $f(t-t_0)U(t-t_0)$.

4.10 Prove the initial value theorem, Equation 4.26.

4.11 Find $\pounds[t^{1/2}]$ and $\pounds[t^{-1/2}]$.

4.12 The methodology of Section 4.4.4 ignores the possibility of repeated conjugate pairs. Remedy this omission and test your remedy by inverting

$$\frac{s^3}{(s^2+1)^2}$$

# CHAPTER 5

# LAPLACE TRANSFORM SOLUTIONS TO DIFFERENTIAL EQUATIONS

Chapter 4 provided the theoretical background to solve initial value problems by the Laplace transform technique. The conceptual approach is simple. We take the Laplace transform of an ODE term by term. The various initial conditions are applied as part of this process. The result is an algebraic equation containing the transformed dependent variable, $Y = Y(s) = \pounds[y(t)]$. The algebraic equation is solved for Y as a function of s. This result is then inverted to find y(t). Throughout this chapter, we shall assume all ODEs are linear with constant coefficients unless explicitly stated otherwise.

**EXAMPLE 5.1** Use Laplace transforms to solve

$$y'' + 4y' + 3y = e^{-t}$$

subject to the initial conditions that $y(0+) = 1$, $y'(0+) = 0$.

**Solution:** The linearity property of Laplace transformation allows the ODE to be transformed term by term. The result is

$$s^2Y(s) - sy(0+) - y'(0+) + 4sY(s) - 4y(0+) + 3Y(s) = \frac{1}{s+1}$$

Solving for Y(s) gives

$$Y(s) = \frac{1 + (s+1)\,[sy(0+) + 4y(0+) + y'(0+)]}{(s+1)\,(s^2 + 4s + 3)}$$

Note that the initial conditions on y and y ' appear directly in the expression for Y(s). Using the specified values gives

$$Y(s) = \frac{1 + (s+1)(s+4)}{(s+1)(s^2+4s+3)} = \frac{s^2+5s+5}{(s+1)^2(s+3)}$$

Since a root is repeated, the sum of partial fractions is written as

$$Y(s) = \frac{K_{11}}{(s+1)} + \frac{K_{12}}{(s+1)^2} + \frac{K_2}{(s+3)}$$

Collecting terms over the common denominator gives

$$\frac{K_{11}(s+1)(s+3) + K_{12}(s+3) + K_2(s+1)^2}{(s+1)^2(s+3)} = \frac{s^2+5s+5}{(s+1)^2(s+3)}$$

The K are found by equating coefficients of powers of s in the numerator:

For $s^2$, $K_{11} + K_2 = 1$

For s, $K_{12} + 4K_{11} + 2K_2 = 5$

For $s^0$, $3K_{12} + 3K_{11} + K_2 = 5$

The solution to this set of simultaneous, algebraic equations is $K_{11} = 5/4$, $K_{12} = 1/2$, $K_2 = -1/4$. Thus,

$$Y(s) = \frac{1/2}{(s+1)^2} + \frac{5/4}{s+1} - \frac{1/4}{s+3}$$

The inverse is

$$y(t) = \frac{1}{2}te^{-t} + 5/4\,e^{-t} - 1/4\,e^{-3t} = \frac{1}{4}[(2t+5)e^{-t} - e^{-3t}]$$

Verification: Differentiate y(t) to obtain

$$y'(t) = -3/4\,e^{-t} - 1/2\,te^{-t} + 3/4\,e^{-3t}$$

$$y''(t) = 1/4\,e^{-t} + 1/2\,te^{-t} - 9/4\,e^{-3t}$$

Substituting back into the original ODE gives

$$\left[\frac{1}{4}e^{-t}+\frac{1}{2}te^{-t}-\frac{9}{4}e^{-3t}\right]+4\left[-\frac{3}{4}e^{-t}-\frac{1}{2}te^{-t}+\frac{3}{4}e^{-3t}\right]$$

$$+3\left[\frac{5}{4}e^{-t}+\frac{1}{2}te^{-t}-\frac{1}{4}e^{-3t}\right]=e^{-t}$$

Substitution of $t = 0$ into $y(t)$ and $y'(t)$ gives $y(0+) = 1$ and $y'(0+) = 0$. The ODE is satisfied. The initial conditions are satisfied. Our result for $y(t)$ is thus confirmed.

---

Example 5.1 illustrates the application of Laplace transform techniques to solving an initial value problem. A great many examples could be given, and many will in fact be included in this chapter and in the problems at the end. However, we turn now to using the solutions to understand how certain physical systems behave in response to inputs. We begin by studying the simplest dynamic system: one governed by a single, linear ODE with constant coefficients.

## 5.1 FIRST ORDER SYSTEMS

Consider the following ODE:

$$V\frac{da}{dt} + Qa = Qa_{in}(t) \tag{5.1}$$

where V and Q are constant. The chemical engineer will recognize Equation 5.1 as governing the output of nonreactive tracer from a stirred tank subject to an input of $a_{in}(t)$. Using different notation,

$$L\frac{di}{dt} + Ri = v(t) \tag{5.2}$$

The electrical engineer will recognize Equation 5.2 as governing the current flow in an RL circuit. The mechanical engineer might write

$$M\frac{du}{dt} + Du = f(t) \tag{5.3}$$

to describe the velocity of a body subject to frictional resistance and a body force f(t). As written, the quantities in these equations are different physically with different physical units. However, they can all be put in a common form:

$$\bar{t}\frac{da}{dt} + a = a_{in}(t) \tag{5.4}$$

where $\bar{t}$ is the **time constant** for the system: $\bar{t} = V/Q$ for Equation 5.1; $\bar{t} = L/R$ for Equation 5.2; and $\bar{t} = M/D$ for Equation 5.3. Thus, the various equations can be made to look alike although the quantities a and $a_{in}$ will still have different units depending on the application. We can get a dimensionless version of Equation 5.4 by scaling:

$$y = a/a_0 \qquad \tau = t/\bar{t} \tag{5.5}$$

so that the ODE becomes

$$\frac{dy}{d\tau} + y = y_{in}(\tau) \tag{5.6}$$

Equation 5.6 is now identical for all the example systems. The solution to Equation 5.6 subject to various initial conditions y(0+) will apply to most systems governed by single, first order ODEs. It should be recognized that Equation 5.6 is not the most general first order, linear ODE. The form

$$\frac{dy}{d\tau} + p(\tau)y = y_{in}(\tau) \tag{5.7}$$

which has a time varying coefficient, is more general. Also, the solutions to

$$\frac{dy}{d\tau} - y = y_{in}(\tau) \tag{5.8}$$

are quite different from the solutions to Equation 5.6. Referring back to the characteristic equations of Chapter 3, Equation 5.6 has a root of s = -1, and Equation 5.8 has a root of s = +1. To avoid the need for separate equations, we will postpone time scaling until after an analytical solution has been obtained. Also, we may wish to consider cases where $a_0 = a(0+) = 0$ so that concentration scaling, at least by $a_0$, is inappropriate. Thus, we will use the following unscaled form for a first order, linear ODE with constant coefficients:

$$\frac{da}{dt} + \alpha a = \alpha a_{in}(t) \tag{5.9}$$

where $\alpha$ has units of $[\text{t-units}]^{-1}$ and may be either positive or negative. Note that a and $a_{in}$ have identical units. The forcing function, $f(t) = \alpha\, a_{in}(t)$, has units of [a-units]/[t-units].

We now consider the response of a first order system to a variety of standard inputs. The Laplace transform of Equation 5.9 is

$$sA(s) - a(0+) + \alpha A(s) = \alpha A_{in}(s)$$

or

$$A(s) = \frac{\alpha A_{in}(s) + a_0}{s + \alpha} \tag{5.10}$$

where $a_0 = a(0+)$.

In the Laplace domain, Equation 5.10 is the general solution to the first order initial value problem (at least for a linear ODE with constant coefficients). We now consider the form of a(t) that corresponds to certain standard inputs and initial conditions.

## 5.1.1 Response to Zero Input

This case, known as the **zero input** response, has $a_{in} = 0$ for all $t > 0$. Equation 5.10 becomes

$$A(s) = \frac{a_0}{s + \alpha} \tag{5.11}$$

Since $a_0$ is a constant, independent of s, Equation 5.11 can be inverted to give

$$a(t) = a_0 e^{-t/\bar{t}} \tag{5.12}$$

This is simple, **exponential decay** for the normal case of $t > 0$. (It is an exponential increase if $t < 0$.) The scaled version of Equation 5.12 is

$$y = e^{-\tau} \tag{5.13}$$

which is the universal equation for exponential decay. The physical interpretation is quite clear. Tracer is gradually purged from the stirred tank; the current in the RL circuit gradually declines; the moving mass gradually slows.

An initially relaxed first order system has $a_0 = a(0+) = 0$. The response of an initially relaxed system is termed its **zero state response**. If the input to an initially relaxed system is zero, the system will remain relaxed. The zero state response to zero input is zero in any linear system. (But do read Section 11.4 on stability. Small inputs, even noise, can grow if the system is unstable.)

A system approaches a steady state if a(t) becomes constant and a' becomes zero as $t \to \infty$. The **steady state response** is denoted as $a(\infty)$, provided the steady state exists. Not all systems achieve a steady state. Even a first order system will not have a steady state if the forcing function, $a_{in}(t)$, is periodic in time. The value for $a(\infty)$ can be determined from the time domain response by taking the limit as $t \to \infty$. It can also be found from the Laplace domain response using the final value theorem in Table 4.2.

**EXAMPLE 5.2** Use the final value theorem to determine $a(\infty)$ for a first order system with zero input.

**Solution:** The final value theorem states

$$a(\infty) = \lim_{t \to \infty} a(t) = \lim_{s \to 0} sA(s)$$

Equation 5.11 gives

$$a(\infty) = \lim_{s \to 0} \frac{s\, a_0}{s + \alpha} = \frac{0}{\alpha} = 0$$

There is no indeterminacy in this result. However, if $\alpha = 0$, Equation 5.9 reduces to

$$\frac{da}{dt} = 0$$

so that $a(\infty) = a_0$ for this special case. For $\alpha < 0$, the final value theorem fails. It fails whenever A(s) has poles with positive real parts. In such situations, the output, a(t), will grow without bounds and $a(\infty) = \pm\infty$.

---

The **transient response** of a system is defined as its total response minus its steady state response, $a(t) - a(\infty)$. For the zero input case considered above, the transient response and total response are identical since $a(\infty) = 0$.

## 5.1.2 Step Input Response

In this and the next few sections, we will determine the response of a first order system to various forcing functions, $a_{in}(t)$. We first consider the step input response. Suppose $a_{in}(t) = U(t)$ so that the input corresponds to a unit step change at $t = 0$. For $a_{in} = U(t)$, $A(s) = 1/s$. Equation 5.10 becomes

$$A(s) = \frac{(\alpha/s) + a_0}{s + \alpha} = \frac{\alpha + s\, a_0}{s\,(s + \alpha)} \qquad (5.14)$$

Equation 5.14 is the step input response in the Laplace domain. It is now inverted to find a(t), which is the time domain response to a step input. Expand Equation 5.14 as

$$A(s) = \frac{1}{s} + \frac{a_0 - 1}{s + \alpha}$$

and invert to give

$$a(t) = U(t) + (a_0 - 1)e^{-\alpha t} \qquad (5.15)$$

When the system is initially relaxed,

$$a(t) = U(t) - e^{-\alpha t} \qquad (5.16)$$

This can be written equally well as

$$a(t) = 1 - e^{-\alpha t}, \qquad t > 0 \qquad (5.17)$$

Note that a(t) must be dimensionless since the right-hand side of this equation is dimensionless. This situation was forced by assuming $a_{in}(t) = U(t)$ since U(t) is dimensionless. A more general input function is $a_{in}(t) = \hat{a}U(t)$ where $\hat{a}$ has the same dimensions as $a_{in}(t)$ and a(t). Instead of Equation 5.17 we would have

$$\frac{a_{in}(t)}{\hat{a}} = 1 - e^{-\alpha t}, \qquad t > 0 \qquad (5.18)$$

As written, Equation 5.17 assumes a(t) and $a_{in}(t)$ are dimensionless and sets $\hat{a} = 1$.

**EXAMPLE 5.3** Apply the final value theorem to Equation 5.14

**Solution:**

$$a(\infty) = \lim_{s \to 0} \frac{s(\alpha + s\,a_0)}{s\,(s + \alpha)} = 1$$

so that the steady state response is always 1 regardless of $a_0$. Since the input is 1, the output of a first order system must eventually approach 1 for any $a_0$.

---

### 5.1.3 Square Wave Response

The function $a_{in}(t) = U(t) - U(t\text{-}t_0)$ is a square wave of amplitude 1 and duration $t_0$:

$$
\begin{aligned}
&U(t) - U(t\text{-}t_0) = 0 && t<0 \\
&U(t) - U(t\text{-}t_0) = 1 && 0<t<t_0 \qquad (5.19)\\
&U(t) - U(t\text{-}t_0) = 0 && t>t_0
\end{aligned}
$$

Its Laplace transform is

$$A_{in}(s) = \pounds[U(t) - U(t-t_0)] = \frac{1}{s} - \frac{e^{-t_0 s}}{s}$$

where we have used the time shifting property in Table 4.2 with $f(t) = 1$ to determine $\pounds[U(t\text{-}t_0)]$. The Laplace domain response for an initially relaxed system follows from Equation 5.10 with $a_0 = 0$:

$$A(s) = \frac{\alpha}{s(s+\alpha)} - \frac{\alpha e^{-t_0 s}}{s(s+\alpha)} \qquad (5.20)$$

Expand these as

$$A(s) = \frac{1}{s} - \frac{1}{s+\alpha} - \frac{e^{-t_0 s}}{s} + \frac{e^{-t_0 s}}{s+\alpha}$$

The first two terms are readily inverted as in Equation 5.15. The next two terms have the same form except for the factor $e^{-t_0 s}$. This corresponds to a time shift, $t \to t\text{-}t_0$, and the time shift theorem in Table 4.2 can again be used. The result, term by term, is

$$a(t) = U(t) - e^{-\alpha t} - U(t\text{-}t_0) + U(t\text{-}t_0)e^{-\alpha(t-t_0)} \qquad (5.21)$$

This result could have been written equally well as

$$a(t) = 1 - e^{-\alpha t} - U(t\text{-}t_0) + U(t\text{-}t_0)e^{-\alpha(t-t_0)}$$

or

$$a(t) = U(t) - U(t)e^{-\alpha t} - U(t\text{-}t_0) + U(t\text{-}t_0)e^{\alpha(t-t_0)}$$

Since all our results are restricted to $t > 0$ and since $U(t) = 1$ for $t > 0$, factors of $U(t)$ can be added or deleted as desired for the sake of appearances. $U(t-t_0)$ must be treated more carefully since $U(t-t_0) = 0$ for $t < t_0$. This time interval includes some portion of positive time, $t > 0$. To write $a(t)$ in more conventional notation we must use two separate expressions:

$$a(t) = 1 - e^{-\alpha t}, \qquad t < t_0 \tag{5.22a}$$

$$a(t) = -e^{-\alpha t} + e^{-\alpha(t-t_0)}, \qquad t > t_0 \tag{5.22b}$$

The first of these, Equation 5.22a, is the same as the response to a unit step input, Equation 5.17, at least until $t = t_0$. This is as it should be since a square wave is a positive step followed by a negative step, and the negative step has not yet occurred in Equation 5.22a.

Equation 5.22b can be written as

$$a(t) = [1-e^{-at_0}]\, e^{-\alpha(t-t_0)} = a(t_0)e^{-a(t-t_0)}, \qquad t > t_0$$

We can now substitute $T = t - t_0$ to obtain

$$a(T) = a(T = 0)\, e^{-\alpha T}, \qquad t > 0$$

which is the same as a zero input response, Equation 5.12, starting at time $T = 0$. This is where the negative step occurred and the input went to zero.

### 5.1.4 Impulse Response

Consider now $a_{in}(t) = \delta(t)$. Equation 5.10 with $a_o = a(0+) = 0$ gives

$$A(s) = \frac{\alpha}{s + \alpha} \tag{5.23}$$

so that

$$a(t) = \alpha e^{-\alpha t} \tag{5.24}$$

The response of an initially relaxed system to a delta function gives the **impulse response function**.

**EXAMPLE 5.4** Verify that Equation 5.24 is the solution to

$$\frac{da}{dt} + \alpha a = \alpha\delta(t) \tag{i}$$

subject to the initial condition that a(0+) = 0.

**Solution:** We must show that Equation 5.24 satisfies the ODE, Equation i, and the initial condition. However, direct substitution into Equation i gives $0 = \delta(t)$. Also, Equation 5.24 gives a(0+) = 1. Thus, neither the ODE nor the initial condition seems to be satisfied.

The impulse response function cannot be directly verified by conventional means. The problem is caused by the infinite discontinuity in $\delta(t)$ at $t = 0$. Considerable insight can be gained by replacing $\delta(t)$ with a finite pulse, D(t), as in Example 4.3. Thus, we seek the solution of

$$\frac{da}{dt} + \alpha a = \alpha D(t) \qquad \text{(ii)}$$

where $D(t) = 1/t_0$, $0 < t < t_0$ and D(t) = 0 elsewhere. We can actually use Laplace transforms to solve this equation. The technique is similar to that used for the square wave input, but now

$$a_{in}(t) = D(t) = \frac{1}{t_0}[U(t) - U(t-t_0)]$$

so that the pulse area is 1 for any $t_0$. The solution of Equation ii is

$$a(t) = \frac{1 - e^{-\alpha t}}{t_0}, \qquad t < t_0 \qquad \text{(iii)}$$

$$a(t) = \frac{-e^{-\alpha t} + e^{-\alpha(t-t_0)}}{t_0}, \quad t > t_0 \qquad \text{(iv)}$$

This result can be verified by conventional means. For any $t_0 > 0$, it satisfies Equation ii and gives a(0+) = 0. Now consider the limit as $t_0 \to 0$.Taking the limit of Equation iv is the easier part. The limit gives an indeterminate form of the 0/0 type. L'Hospital's rule gives

$$\lim_{t_0 \to 0} a(t) = \lim \frac{-0 + \alpha e^{-\alpha(t-t_0)}}{1} = \alpha e^{-\alpha t}$$

so that Equation iv approaches Equation 5.24 as $t_0 \to 0$. Turning to Equation iii, it is true that a = 0 at t = 0 for any $t_0 > 0$, and we might suppose this defines a(0+) even in the limit. However, the actual facts are a bit more subtle. Consider the value of $a(t_0)$ as a function of $t_0$:

$$a(t_0) = \frac{1 - e^{-\alpha t_0}}{t_0}$$

This function has value 0 when $t_0 = \infty$ and value 1 when $t_0 = 0$. The difference between a(0+) and $a(t_0)$ grows larger as the difference between $t_0$ and 0+ grows smaller. What happens is that a(t) has a discontinuity at t = 0 when the input is a delta function. The system remains initially relaxed with a(0-) = 0, but the impulse function causes an instantaneous change to a(0+) = 1. When the input function is an ordinary, piecewise continuous function, this cannot happen. Ordinarily, the output changes only gradually, not instantaneously.

---

Idealized inputs such as unit step changes, square waves, and delta function inputs can be approximated in real systems. To understand their physical significance, consider the stirred tank reactor described by Equation 5.1. Suppose the chemical tracer is salt. The stirred tank is initially relaxed if it is full of fresh water at t = 0. A unit step change is achieved by switching the input stream from fresh water to salt water. The water within the tank and in the outlet stream will gradually grow salty with the salt concentration, a(t), following Equation 5.17. A square wave input is achieved just by shifting the input back to fresh water after some time $t_0$.

For a delta function input, we suppose the tank is initially full of fresh water but at time t = 0, someone throws in a box of salt. Ignoring dissolution time, the tank becomes instantly salty; but the feed to the tank remains fresh; and the salt immediately begins to purge from the system. The outlet salt concentration will follow Equation 5.24 provided the units are properly normalized.

Note that $\delta(t)$ has units of time$^{-1}$ so that the assumed input function, $a_{in}(t) = \delta(t)$, makes no sense when $a_{in}$ is a chemical concentration. The actual input must be

$$a_{in}(t) = \frac{S\delta(t)}{Q} \tag{5.25}$$

where S is the quantity (mass or moles) of salt added. With this input, Equation 5.23 gives

$$A(s) = \frac{S\alpha}{Q(s + \alpha)} \tag{5.26}$$

and the time domain response is

$$a(t) = \frac{S\alpha e^{-\alpha t}}{Q} = \frac{Se^{-\alpha t}}{V} \tag{5.27}$$

where we have used $\alpha = Q/V$ for a stirred tank. Equation 5.27 now makes physical sense; S/V is the concentration in the tank just after the box of salt has been added. The outlet concentration declines exponentially for $t > 0$. Note that Equation 5.27 has the same form as Equation 5.12. Thus, for $t > 0$ we cannot tell the difference between a system that had initial concentration S/V and one that had an initial concentration of zero and received a delta function input. A linear system has no memory.

### 5.1.5 Frequency Response

The final, standard input to be considered is a sine wave, $a_{in}(t) = \sin \omega t$. For an initially relaxed system, Equation 5.10 gives

$$A(s) = \frac{\alpha\,\omega}{(s+\alpha)\,(s^2+\omega^2)} = \left[\frac{K_1}{s+\alpha} + \frac{(K_2 + K_3 s)}{s^2+\omega^2}\right] \qquad (5.28)$$

where $K_1$, $K_2$, and $K_3$ are given by

$$K_1\omega^2 + K_2\alpha = \alpha\omega$$

$$K_2 + K_3\,\alpha = 0$$

$$K_1 + K_3 = 0$$

Solving for $K_1$, $K_2$, and $K_3$ and substituting back into Equation 5.28 gives

$$A(s) = \frac{\alpha\,\omega}{\alpha^2+\omega^2}\left[\frac{1}{s+\alpha} + \frac{\alpha - s}{s^2+\omega^2}\right]$$

$$= \frac{\alpha\,\omega}{\alpha^2+\omega^2}\left[\frac{1}{s+1} + \frac{\alpha}{s^2+\omega^2} - \frac{s}{s^2+\omega^2}\right]$$

Inverting term by term gives

$$a(t) = \frac{\alpha\,\omega}{\alpha^2+\omega^2}\left[e^{-\alpha t} + \frac{\alpha \sin \omega t}{\omega} - \cos \omega t\right]$$

$$(5.29)$$

$$= \frac{\alpha[\omega e^{-\alpha t} + \alpha \sin \omega t - \omega \cos \omega t]}{\alpha^2+\omega^2}$$

This result can be verified by differentiation and back substitution into the governing ODE. We also note that a(0+) = 0 in agreement with the initial condition of an initially relaxed system.

Equivalent to Equation 5.29 is

$$a(t) = \alpha \left[ \frac{\omega e^{-\alpha t} + \sqrt{\alpha^2 + \omega^2}\ \sin(\omega t - \phi)}{(\alpha^2 + \omega^2)} \right] \tag{5.30}$$

where $\phi = \tan^{-1}(\omega/\alpha)$. This combines sine and cosine terms into a single sinewave with a phase shift.

**EXAMPLE 5.5** Verify that Equations 5.29 and 5.30 are equivalent.

**Solution:** This is an exercise in trigonometry. We need to show that

$$\sqrt{\alpha^2 + \omega^2}\ \sin(\omega t - \phi) = \alpha \sin \omega t - \omega \cos \omega t \tag{i}$$

We start with the difference formula

$$\sin(\theta_1 - \theta_2) = \sin\theta_1 \cos\theta_2 - \cos\theta_1 \sin\theta_2 = \cos\theta_2 [\sin\theta_1 - \cos\theta_1 \tan\theta_2]$$

Set $\theta_2 = \tan^{-1}(\omega/\alpha)$. Then $\cos\theta_2 = \cos(\tan^{-1}\omega/\alpha) = \alpha/\sqrt{\alpha^2+\omega^2}$. Now set $\phi = \theta_2$ and $\omega t = \theta_1$. This gives

$$\sin(\omega t - \phi) = \frac{\alpha}{\sqrt{\alpha^2 + \omega^2}} \left[ \sin \omega\tau - \frac{\omega}{\alpha} \cos \omega\tau \right]$$

The factor of $\alpha/\sqrt{\alpha^2+\omega^2}$ is transposed to obtain Equation i.

---

The sinusoidal input is converted to a sinusoidal output with a phase shift of $-\phi$ and an amplitude dampening of $\alpha/\sqrt{\alpha^2+\omega^2}$. The transient response corresponds to the $\exp(-\alpha t)$ term in Equation 5.30. This term dies out as $t \to \infty$, but the sinusoidal output continues. The system does not approach a true steady state with a' = 0. The final value theorem falsely predicts $a(\infty) = 0$. It does gives the correct time average output,

$$\bar{a} = \lim_{T \to \infty} \frac{1}{T} \int_0^T a(t)dt$$

when $a_{in}(t)$ is periodic.

## 5.2 SECOND ORDER SYSTEMS

Corresponding to Equation 5.4, the general equation for a linear, second order system with constant coefficients is

$$t_1 t_2 \frac{d^2a}{dt^2} + (t_1 + t_2)\frac{da}{dt} + a = a_{in}(t) \tag{5.31}$$

where $t_1$ and $t_2$ are the two time constants for the system. These time constants may be real or imaginary. If imaginary, they will be conjugate pairs so that the coefficients of Equation 5.31 will be real.

It is possible to convert Equation 5.31 to dimensionless form with time scaling based on either $t_1$ or $t_2$. Suppose we scale using $t_1$. Then

$$K\frac{d^2y}{d\tau^2} + (1 + K)\frac{dy}{d\tau} + y = y_{in}(t) \tag{5.32}$$

where $K = t_2/t_1$. However, this presupposes that $t_1$ is real and positive. Special forms of Equation 5.32 are needed when $t_1$ is negative or imaginary. Scaling is best done after an analytical solution has been obtained.

An alternative form to Equation 5.31 is

$$\frac{d^2a}{dt^2} + (\alpha + \beta)\frac{da}{dt} + \alpha\beta a = \alpha\beta a_{in}(t) \tag{5.33}$$

where $\alpha = 1/t_1$ and $\beta = 1/t_2$. The Laplace transform of Equation 5.33 is

$$s^2A(s) - sa(0+) - a'(0+) + (\alpha + \beta)\,[sA(s) - a(0+)]$$

$$+ \alpha\beta A(s) = \alpha\beta A_{in}(s)$$

Solving for A(s) gives the Laplace domain solution to the second order initial value problem

$$A(s) = \frac{\alpha\beta A_{in}(s) + (\alpha+\beta+s)\, a(0+) + a'(0+)}{s^2 + (\alpha + \beta)s + \alpha\beta}$$

$$= \frac{\alpha\beta A_{in}(s) + (\alpha+\beta+s)\, a(0+) + a'(0+)}{(s + \alpha)\,(s + \beta)} \tag{5.34}$$

To complete the solution, we assign initial conditions for a and a' and determine the Laplace transform of $a_{in}(t)$. Equation 5.34 is then inverted to find a(t). Note that $(s + \alpha)(s + \beta) = 0$ is the **characteristic equation** for the second order system under consideration. Compare Equation 3.7. The roots are $s_1 = -\alpha$ and $s_2 = -\beta$. There are three different cases to be considered:

(i) Paired real roots, $\alpha = \beta$.

(ii) Distinct real roots, $\alpha \neq \beta$.

(iii) Conjugate imaginary roots, $\alpha, \beta = \sigma \pm i\omega$.

The zero input response, step input response, square wave response, impulse response, and frequency response can be found for each of these cases with arbitrary a(0+) and a'(0+). The impulse response of an initially relaxed system will serve as a model.

## 5.2.1 Impulse Response

This case has $a(0+) = a'(0+) = 0$ and $A_{in}(s) = 1$. Equation 5.34 becomes

$$A(s) = \frac{\alpha\beta}{(s+\alpha)\,(s+\beta)} \tag{5.35}$$

The various cases are

(i) Paired real roots

$$A(s) = \frac{\alpha\beta}{(s+\alpha)^2}, \qquad a(t) = \alpha^2 t e^{-\alpha t} \tag{5.36}$$

(ii) Distinct real roots

$$A(s) = \frac{\alpha\beta}{(s+\alpha)\,(s+\beta)}, \quad a(t) = \frac{\alpha\beta(e^{-\alpha t} - e^{-\beta t})}{\beta - \alpha} \tag{5.37}$$

(iii) Conjugate imaginary roots

$$A(s) = \frac{\sigma^2 + \omega^2}{(s+\sigma)^2 + \omega^2}, \quad a(t) = \left(\frac{\sigma^2 + \omega^2}{\omega}\right) e^{-\sigma t} \sin \omega t \tag{5.38}$$

where $\alpha = \sigma + i\omega$, $\beta = \sigma - i\omega$, $\alpha + \beta = 2\sigma$, and $\alpha\beta = \sigma^2 + \omega^2$.

It can be argued that Equation 5.37 includes the other two cases. Equation 5.38 can be derived from it using the identities

$$\sin x = \frac{e^{ix} - e^{-ix}}{2i}, \quad \cos x = \frac{e^{ix} + e^{-ix}}{2}$$

Equation 5.36 can be derived from it by taking the limit as $\beta \to \alpha$ with $\alpha$ held constant. Equation 5.36 can also be derived from Equation 5.38 by taking the limit as $\omega \to 0$ so that the conjugate pair becomes a real pair with $\alpha = \beta = \sigma$.

**EXAMPLE 5.6** Determine the step change response for an initially relaxed, second order system having paired real roots.

**Solution:** For $a_{in}(t) = U(t)$, $A_{in}(s) = 1/s$. Thus, we need to invert

$$A(s) = \frac{\alpha^2}{s(s+\alpha)^2} \qquad \text{(i)}$$

This can be done in the usual manner using partial fraction expansions:

$$A(s) = \frac{K_0}{s} + \frac{K_1}{s+\alpha} + \frac{K_2}{(s+\alpha)^2} \qquad \text{(ii)}$$

An alternative approach is to note that division by s is equivalent to integration with respect to t. Thus,

$$a(t) = \int_0^t \alpha^2 t\, e^{-\alpha t}\, dt = -e^{-\alpha t}(\alpha t + 1) \Big|_0^t$$
$$= 1 - (1 + \alpha t)\, e^{-\alpha t} \qquad \text{(iii)}$$

where the integrand in Equation iii was obtained from Equation 5.36. Taking the Laplace transform of Equation iii gives

$$A(s) = \frac{1}{s} - \frac{1}{s+\alpha} - \frac{\alpha}{(s+\alpha)^2} = \frac{\alpha^2}{s(s+\alpha)^2}$$

which is seen to agree with Equations i and ii.

---

## 5.3 THIRD AND HIGHER ORDER SYSTEMS

The linear, third order equation with constant coefficients is

$$\frac{d^3a}{dt^3} + (\alpha+\beta+\gamma)\frac{d^2a}{dt^2} + (\alpha\beta+\alpha\gamma+\beta\gamma)\frac{da}{dt} + \alpha\beta\gamma a = \alpha\beta\gamma a_{in}(t) \qquad (5.39)$$

This equation has either three real roots (with possible repetition) or one real root and a conjugate pair of complex roots. The Laplace transform for an initially relaxed, third order system is

$$A(s) = \frac{\alpha\beta\gamma \, A_{in}(s)}{(s+\alpha)\,(s+\beta)\,(s+\gamma)} \tag{5.40}$$

For the case where there is only one real root, Equation 5.40 can be written as

$$A(s) = \frac{\alpha\beta\gamma \, A_{in}(s)}{(s+\alpha)\,[(s+\sigma)^2+\omega^2]} \tag{5.41}$$

Inversion proceeds by the usual method of partial fraction expansion.

**EXAMPLE 5.7** Find the solution to the following initially relaxed third order system:

$$a''' + 4a'' + 5a' + 2a = 8\delta(t)$$

**Solution:** The first problem is to factor the characteristic equation

$$s^3 + 4s^2 + 5s + 2 = 0$$

Analytical factorization is always possible for algebraic equations up to fourth order, but can be a chore. Here, the factors are s = 1 (repeated) and s = 2. Equation 5.40 gives

$$A(s) = \frac{8}{(s+1)^2\,(s+2)} = \frac{K_0}{s+2} + \frac{K_1}{(s+1)} + \frac{K_2}{(s+1)^2}$$

The direct algebraic approach for finding the coefficients becomes tedious, but $K_0$ can quickly be found using

$$K_0 = (s+2)\,A(s)\,\Big|_{s=-2} = \frac{8}{(-2+1)^2} = 8$$

Similarly, we can find $K_2$ from

$$K_2 = (s+1)^2\,A(s)\Big|_{s=-1} = 8$$

This trick of multiplying by $(s\text{-}s_j)$ and then setting $s = s_j$ will not work for $K_1$. Equation 4.41 could be used to find $K_1$, but with $K_0$ and $K_2$ known, we can also revert to the algebraic approach

$$A(s) = \frac{8(s+1)^2 + K_1\,(s+1)\,(s+2) + 8(s+2)}{(s+1)^2\,(s+2)} = \frac{8}{(s+1)^2\,(s+2)}$$

from which we quickly obtain $K_1 = -8$. Thus,

$$A(s) = \frac{8}{s+2} - \frac{8}{(s+1)} + \frac{8}{(s+1)^2}$$

and

$$a(t) = 8e^{-2t} - 8e^{-t} + 8te^{-t}$$

$$= 8e^{-t}\,[e^{-t} - 1 + t]$$

---

Fourth and higher order equations present no conceptual difficulties, but factoring the characteristic equation can become difficult or even impossible except numerically. Solving for the $K_j$ in a partial fraction expansion is easy when $s_j$ is unrepeated. It quickly becomes difficult for highly repeated roots. As a practical matter, there are very few cases where higher than fourth order systems need to be solved analytically.

## 5.4 SIMULTANEOUS, FIRST ORDER ODEs

The canonical problem in this book is a set of simultaneous, first order ODEs. When they are all linear with constant coefficients, Laplace transform techniques can be used for their solution. The equations are

$$\begin{aligned}
\frac{da}{dt} + \alpha_A a + \beta_A b + \gamma_A c + \cdots &= f_A \\
\frac{db}{dt} + \alpha_B a + \beta_B b + \gamma_B c + \cdots &= f_B \qquad (5.42) \\
\frac{dc}{dt} + \alpha_C a + \beta_C b + \gamma_C c + \cdots &= f_C \\
\vdots \qquad\qquad &\quad \vdots
\end{aligned}$$

Define $A(s) = \pounds[a(t)]$, $B(s) = \pounds[b(t)]$, $C(s) = \pounds[c(t)]$, and so on. Then Equation 5.42 can be transformed to give

$$sA(s) - a(0+) + \alpha_A A(s) + \beta_A B(s) + \gamma_A C(s) + \ldots = F_A(s)$$

$$sB(s) - b(0+) + \alpha_B A(s) + \beta_B B(s) + \gamma_B C(s) + \ldots = F_B(s) \qquad (5.43)$$

$$sC(s) - c(0+) + \alpha_C A(s) + \beta_C B(s) + \gamma_C C(s) + \ldots = F_C(s)$$

This is a set of simultaneous, algebraic equations in the unknowns A(s), B(s), C(s), ... . When solved, the results can be inverted to give a(t), b(t), c(t), ... . The reader will note that we now have a choice of methods for solving simultaneous ODEs. Chapter 3 showed how Equation 5.42 can be combined to form a single, nth order ODE. Alternatively, Equation 5.43 can be solved algebraically for A(s), B(s), C(s), ... . These results are then inverted to give a(t), b(t), c(t), ... .. Mathematical consistency ensures that either approach gives the same solution.

**EXAMPLE 5.8** Solve the following set of ODEs for an initially relaxed system:

$$a' + a - b = e^{-t} \qquad (i)$$

$$b' + 2a - b = \delta(t) \qquad (ii)$$

**Solution:** The Laplace transforms are

$$sA + A - B = 1/(s+1)$$

$$sB + 2A - B = 1$$

where we have set a(0+) = b(0+) = 0. Solving for A gives

$$A(s) = \frac{2s}{(s+1)(s^2+1)} \qquad (iii)$$

Inversion is left to the reader.

We can also combine Equations i and ii into a second order ODE. The result is

$$a'' + a = \delta(t) - 2e^{-t}$$

Taking the Laplace transform gives

$$s^2A - sa(0+) - a'(0+) + A = 1 - \frac{2}{s+1}$$

Here, a(0+) = 0 since the system is initially relaxed. It is tempting to set a'(0+) = 0 as well. However, this is not inherent in the definition of an initially relaxed system. Instead, Equation i gives a'(0+) = 1. Thus,

$$s^2A - 0 - 1 + A = 1 - \frac{2}{s+1}$$

or

$$A = \frac{1 - \frac{2}{s+1} + 1}{s^2 + 1} = \frac{2s}{(s+1)(s^2+1)}$$

which agrees with our previous result, Equation iii.

---

## 5.5 PARTIAL DIFFERENTIAL EQUATIONS

This book is about lumped parameter systems governed by ODEs, not distributed parameter systems governed by PDEs. However, Laplace transform techniques provide such a useful tool for solving certain PDE's that the method is worth a brief description for future reference. We suppose now that a(t,z) is a function of two independent variables, t and z. We can take the Laplace transform with respect to one of the independent variables and leave the other untransformed. Thus,

$$A(s,z) = \int_0^\infty a(t,z)e^{-st}\,dt \tag{5.44}$$

then

$$\frac{dA}{dz} = \int_0^\infty \frac{\partial a}{\partial z} e^{-st}\,dt \tag{5.45}$$

and in general

$$\frac{d^nA}{dz^n} = \int_0^\infty \frac{\partial^n a}{\partial z^n} e^{-st}\,dt \tag{5.46}$$

Linear PDEs with constant coefficients can be transformed using these relationships. The result is a linear ODE, with constant coefficients, in the transformed variable, A(s,z).

**EXAMPLE 5.9** The governing equation for an unsteady piston flow reactor is

$$\frac{\partial a}{\partial t} + u\frac{\partial a}{\partial z} = R_A \qquad \text{(i)}$$

where u, the fluid velocity, is assumed constant. Solve this equation assuming a first order reaction, $R_A$ = -ka. The initial condition is a(0,z) = 0 so that, initially, there is no reactant in the system. The boundary condition is a(t,0) = $a_0$ so that the inlet concentration is $a_0$ for all t > 0.

**Solution:** We transform Equation i

$$\int_0^\infty \frac{\partial a}{\partial t} e^{-st}\,dt + u\int_0^\infty \frac{\partial a}{\partial z} e^{-st}\,dt = -k\int_0^\infty a\,e^{-st}\,dt$$

The first and last integrals are evaluated as if there were no z dependence. The middle integral is evaluated using Equation 5.45. The result is

$$sA(s,z) - a(0+,z) + u\frac{dA}{dz} = -kA$$

This is a linear ODE in A with a single independent variable, z. Setting a(0+,z) = 0 according to the initial condition gives

$$\frac{dA}{dz} + \left(\frac{k+s}{u}\right)A = 0$$

The solution is

$$A(s,z) = C_0\, e^{-\left(\frac{k+s}{u}\right)z}$$

We know the boundary condition, a(t,0) = 1, for a. This must be transformed to give a boundary condition for A. Transformation gives A(s,0) = 1/s. Thus,

$$A(s,z) = \frac{e^{-(k+s/u)z}}{s} \qquad \text{(ii)}$$

Equation ii is the Laplace transform of the solution a(t,z). To invert, we regard z as a constant. Thus

$$A(s,z) = e^{-kz/u}\,\frac{e^{-st_0}}{s} \qquad \text{(iii)}$$

where $t_0 = z/u$. Equation iii corresponds to a step change of amplitude exp(-kz/u) which has been time-shifted. (See, for example, Section 5.1.3). The solution is

$$a(t,z) = e^{-kz/u}\, U(t-z/u)$$

Equivalently,

$$a(t,z) = e^{-kz/u}, \qquad t > z/u$$
$$= 0, \qquad t < z/u$$

---

The above approach can be used on second and higher order PDEs with constant coefficients. Typically, the resulting A(s,z) will be an irrational function of s that cannot be inverted by the methods given here. For some problems, however, merely finding A(s,z) gives useful information about the system. For example, properties of the residence time distribution can be found without inversion. See Chapter 10.

## PROBLEMS

5.1 Use Laplace transforms to solve $y' - 3y = 0$ subject to the **boundary** condition that $y(1) = 1$.

5.2 Consider an initially relaxed, first order system

$$y' + y = f(t)$$

where f(t) is a triangular impulse function

$$f(t) = H - t/B, \qquad 0 < t < B$$
$$= 0 \text{ elsewhere}$$

a) Find y(t)

b) Set $B = 2/H$ and take the limit as $H \to \infty$. Discuss this result.

5.3 Solve $a'' + 3a' + 2a = \delta(t)$ subject to the initial conditions that $a(0) = 1$, $a'(0) = 0$.

5.4 The response of a system depends on the forcing function. Consider an initially relaxed first order system. Give an example of a forcing function where the response of the first order system is the same as the impulse response of a second order system.

5.5 Find the input, $a_{in}(t)$, such that the output of an initially relaxed, first order system will be a pure sinewave, $a(t) = \sin \omega t$.

5.6 Find the step input response for an initially relaxed, second order system with real, distinct time constants, $t_1$ and $t_2$.

5.7 Find the step input response for an initially relaxed, second order system with complex roots, $s_1, s_2 = \sigma \pm \omega i$. Do not assume $\sigma = 0$.

5.8 Solve the following set of ODEs given the initial conditions that $a(0) = 1$ and $b(0) = 1$:

$$a' + 2a - b = e^{-t}$$
$$b' + 2a - b = \delta(t)$$

5.9 Consider a general, second order system as in Equations 5.33 and 5.34. Under what conditions can the zero input response be identical to the impulse response. Ask the same question for a first order system and a third order system. Generalize your result to an $N^{th}$ order system.

5.10 The proof of the final value theorem in Section 4.3.8 seems general. Why does it fail for A(s) having a positive pole, $\alpha > 0$. Hint: does A(s) exist for $\alpha > s$?

5.11 Suppose a(t,z) is given by the following PDE:

$$\frac{\partial a}{\partial t} + u\frac{\partial a}{\partial z} = D\frac{\partial^2 a}{\partial z^2}$$

subject to

$$a(0,z) = 0 \qquad \text{(Initial Condition)}$$

$$u\delta(t) = u\,a(t,0) - D\frac{\partial a}{\partial z}\bigg|_0 \qquad \text{(Boundary Condition at } z = 0\text{)}$$

$$\frac{\partial a}{\partial z}\bigg|_L = 0 \qquad \text{(Boundary Condition at } z = L\text{)}$$

Reduce this system to an ODE in A(s,z) together with associated boundary conditions. (This set of equations represents the response of the **axial dispersion model** to an inlet delta function. The model has a long history in chemical engineering. See Nauman and Buffham, **Mixing in Continuous Flow Systems**, Wiley, New York, 1983.)

# CHAPTER 6

# LINEAR ELECTRICAL CIRCUITS

Electrical engineers pioneered what is now called linear systems analysis. Particularly, they applied Laplace transform techniques to ordinary differential equations. In deference to them we study electrical circuits and transients therein. Specifically, we study circuits consisting of resistors, capacitors, and inductors plus voltage or current sources. Table 6.1 illustrates the various circuit elements and gives the voltage and current relationships for them. The reader is assumed familiar with basic concepts of electrical circuits from previous courses in physics.

The current, by convention, is assumed to flow from positive voltages to negative. The sign of the voltage drop across a **passive** element (meaning a resistor, capacitor, or inductor) can be determined by the direction of the current flowing through that element. The sign of the voltage across an **active** element (meaning a battery, voltage generator, or current source) is opposite the direction of the current. For elements in series, all have the same current and the voltages are algebraically additive. We illustrate these points in the following examples.

**EXAMPLE 6.1** Determine the current, i(t), in the following circuit.

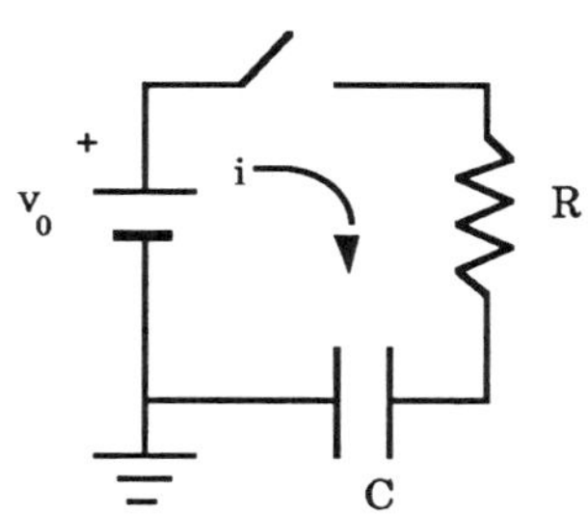

The switch is closed at time $t = 0$. There was no charge on the capacitor at this time.

**Solution:** There are three elements in series. The voltage source, illustrated as a battery in this case, generates a constant voltage of $+v_0$ relative to ground. This must be matched by the voltage drop across the two **passive** elements, the resistor and the capacitor. The voltage drops across these elements are obtained from Table 6.1 as

$$v_R = iR$$

$$v_C = \frac{1}{C}\int_0^t i dt$$

where we have set $v_C(0+) = 0$. Since the resistor and capacitor are connected in series, their combined voltage drop is just $v_R + v_C$. Thus, for $t > 0$,

$$\begin{aligned} v_o &= v_R + v_C \\ &= iR + \frac{1}{C}\int_0^t i dt \end{aligned}$$

This is a linear integral equation that is easily solved by Laplace transform techniques. The transform is

$$\frac{v_o}{s} = I(s)R + \frac{1}{Cs} I(s)$$

or

$$I(s) = \frac{v_o R}{s + \frac{1}{RC}}$$

The inverse is

$$i(t) = (v_o/R)e^{-t/RC}$$

which is the well known formula for exponential decay in an RC circuit. The time constant for the decay is $\bar{t} = (RC)^{-1}$.

**TABLE 6.1**
**VOLTAGE-CURRENT RELATIONSHIPS FOR ELECTRICAL ELEMENTS**

| Element | Symbol | Voltage | Laplace Transform of Voltage | Current | Laplace Transform of Current |
|---|---|---|---|---|---|
| Battery | $v_0$ 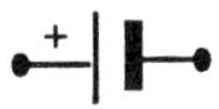 | $v_0$ Independent of current | $\frac{v_0}{s}$ | Unknown, to be determined | Unknown, to be determined |
| Voltage source | v(t) 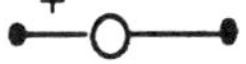 | v(t) Independent of current | V(s) | Unknown, to be determined | Unknown, to be determined |
| Current Source | i(t) 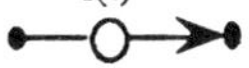 | Unknown, to be determined | Unknown, to be determined | i(t) Independent of voltage | I(s) |
| Resistance | R  $\Leftarrow v_R \Rightarrow$ | $v_R = i_R R$ | $V_R = RI_R$ | $i_R = v_R R$ | $I_R = V_R/R$ |
| Inductance | L  $\Leftarrow v_L \Rightarrow$ | $v_L = L\frac{di_L}{dt}$ | $V_L = sLI_L - Li(0+)$ | $i_L = \frac{1}{L}\int_0^t v_L dt + i_L(0+)$ | $I_L = \frac{V_L}{sL} + \frac{i(0+)}{s}$ |
| Capacitance | C 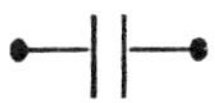 $\Leftarrow v_c \Rightarrow$ | $v_c = \frac{1}{C}\int_0^t i_c\, dt + v_c\ (0+)$ | $V_c = \frac{I_c}{s\,C} + \frac{v\,(0+)}{s}$ | $i_c = C\frac{dv_c}{dt}$ | $I_c = sCV_c - Cv(0+)$ |

**EXAMPLE 6.2** Determine the current, i(t), in the following circuit:

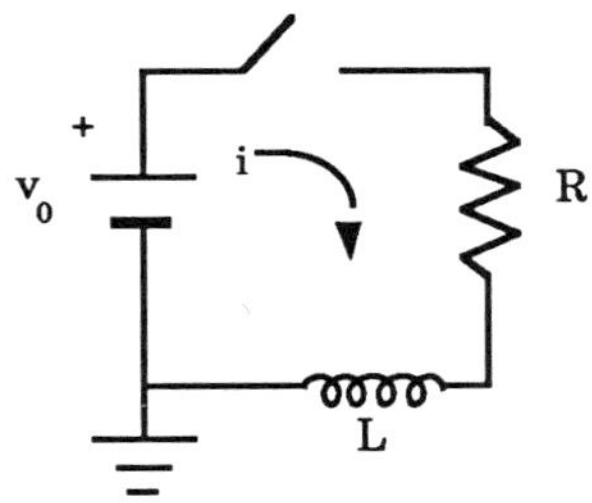


The system is initially relaxed, and the switch is closed at $t = 0$.

**Solution:** This is the same as Example 6.1 except that the capacitor is replaced by an inductor. The result is now an ODE:

$$v_o = v_R(t) + v_C(t)$$
$$= iR + L\frac{di}{dt}$$

or, in the Laplace domain

$$\frac{v_o}{s} = IR + sLI - Li(0+)$$

For the initially relaxed case we have

$$I(s) = \frac{v_o}{s(R+sL)} = \frac{v_o/L}{s(s+R/L)}$$

The inverse is

$$i(t) = (v_o/R)\,(1 - e^{-Rt/L})$$

The RL circuit shows a gradual increase in current flow up to the steady state value of $v_o/R$. Note that the final value theorem in Table 4.2 predicts this steady state value. The time constant for an RL circuit is $\bar{t} = L/R$.

---

**EXAMPLE 6.3** Determine the current, i(t), in the following, initially relaxed circuit:

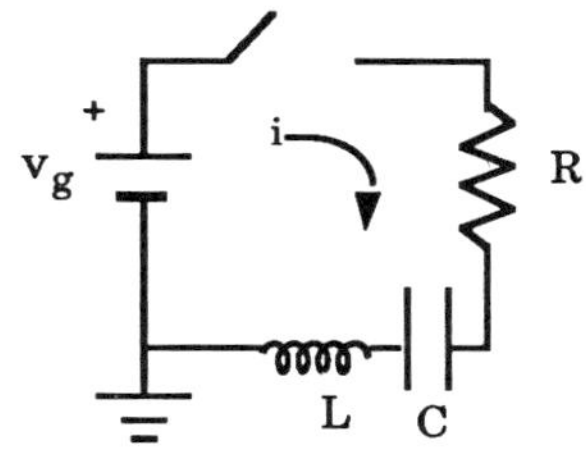


We suppose the switch is closed at time t = 0.

**Solution:** There are now three elements in series and the voltage is summed across them.

$$v_g = v_R(t) + v_C(t) + v_L(t)$$

$$= iR + \frac{1}{C}\int_0^t i dt + L\frac{di}{dt}$$

Differentiation gives a second order ODE.

$$L\frac{d^2i}{dt^2} + R\frac{di}{dt} + \frac{i}{C} = \frac{dv_g}{dt}$$

For the case at hand, $v_g(t) = v_o U(t)$ so that the forcing function, $dv_g/dt$, is a delta function, $v_o\delta(t)$. Taking Laplace transforms gives

$$s^2LI + sRI + \frac{I}{C} = v_o$$

or

$$I = \frac{v_o/L}{s^2 + (R/L)s + \frac{1}{CL}}$$

$$= \frac{v_o/L}{(s+\alpha)(s+\beta)}$$

where

$$\alpha = \frac{R}{2L} + \sqrt{\left(\frac{R}{2L}\right)^2 - \frac{1}{LC}}$$

$$\beta = \frac{R}{2L} - \sqrt{\left(\frac{R}{2L}\right)^2 - \frac{1}{LC}}$$

Heaviside's expansion theorem can now be applied to obtain the inverse, i(t), in the general exponential form

$$i(t) = K_1 e^{-\alpha t} + K_2 e^{-\beta t} \qquad \text{(i)}$$

where $K_1$ and $K_2$ are coefficients to be determined in the usual manner. However, as in our previous dealings with second order systems, this form is most useful for the case where $\alpha$ and $\beta$ are real and distinct. Thus, we use the general form only for the situation where $(R/2L)^2 > 1/LC$ or $R > 2\sqrt{L/C}$. The system is said to be **overdamped** if $R > 2\sqrt{L/C}$. If $R = 2\sqrt{L/C}$ the system has equal real roots and is said to be **critically damped**. In this case, i(t) has the form

$$i(t) = (K_1 + K_2 t)e^{-\frac{Rt}{2L}} \qquad \text{(ii)}$$

The final situation to be considered is where $R < 2\sqrt{L/C}$ so that $\alpha$ and $\beta$ are complex conjugates. This is called the **underdamped** case. The solution for i(t) has the form

$$i(t) = e^{-\frac{Rt}{2L}} [K_1 \sin \omega t + K_2 \cos \omega t] \qquad \text{(iii)}$$

where

$$\omega = \sqrt{1/LC - (R/2L)^2}.$$

---

The series RCL circuit considered in Example 6.3 shows the typical behavior of a **stable**, second order system. The real parts of the roots $\alpha$, $\beta$ are negative so that disturbances are eventually damped. More general behavior for a second order system such as permanent oscillation or runaways can only occur when a root has a real portion which is zero or positive. Graphic illustrations of the response are postponed until Chapter 12.

## 6.1 KIRCHOFF 'S RULES

We now introduce two rules useful for analyzing parallel and complex circuits.

1. The net flow of current into a **node** is zero. A node is simply a connection point (junction, terminal point) in the circuit. Each electrical element has two nodes. Circuits are made by joining nodes.

2. The net voltage difference about any closed circuit is zero.

The first of Kirchoff 's rules represents an electron balance. It states that charge does not accumulate at a node. A typical node is illustrated by

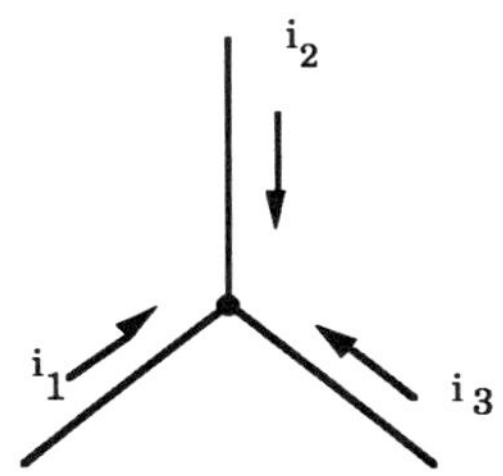


where we have shown the currents as though each were flowing toward the node. For this situation, $i_1 + i_2 + i_3 = 0$. We usually draw the currents in the direction of actual flow. (Recall that current flow, by convention, goes from a positive voltage to a negative, or less positive voltage.) In complex circuits, the actual direction of flow may not be obvious, but an arbitrary guess is satisfactory to begin the analysis. If wrong, the calculated current will just be negative. The following node diagram

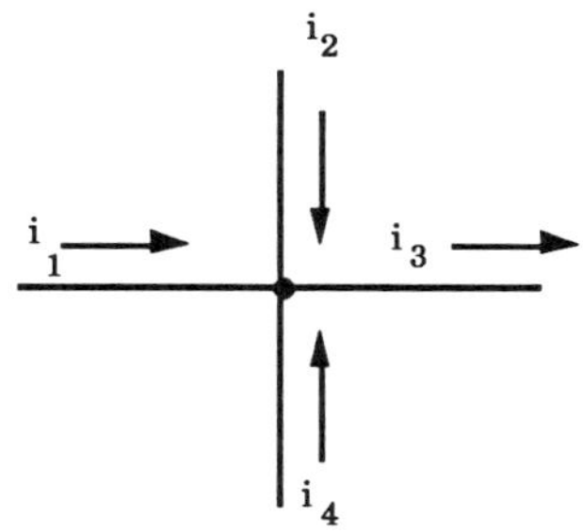


presupposes the direction of current flow. For it, Kirchoff 's rule takes the form $i_1 + i_2 - i_3 + i_4 = 0$.

The second of Kirchoff 's rules can be restated as "The voltage at any node is independent of the path taken to reach that node." Consider the following parallel circuit

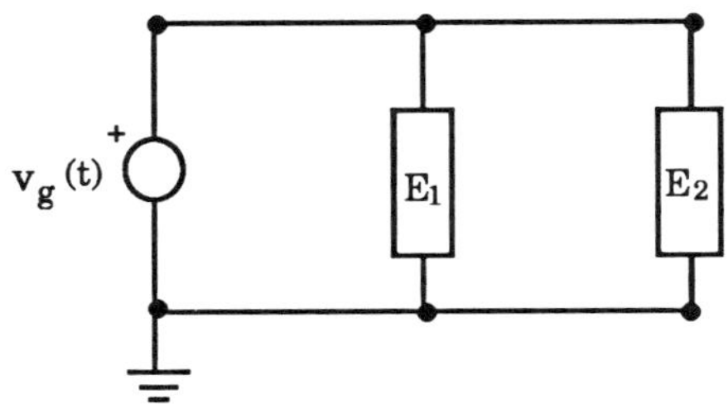


where $E_1$ and $E_2$ are passive elements. As drawn, the circuit has six nodes. However, if we are willing to ignore the impedance of the wiring, the circuit can be redrawn to have only two nodes:

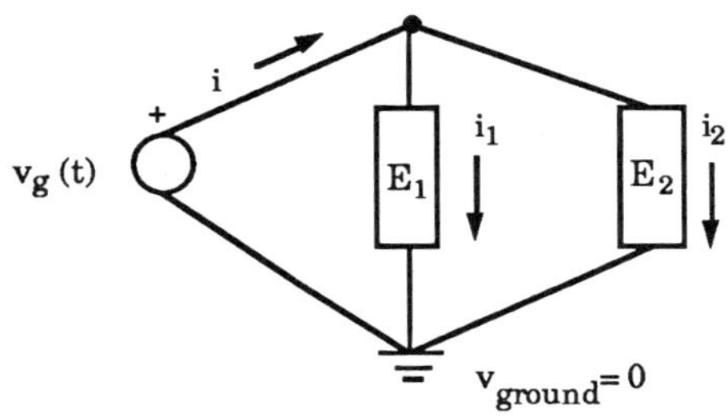


Kirchoff 's first rule gives $i = i_1 + i_2$. The second rule shows that the voltage drop, $v - v_{ground} = v$, is the same for both $E_1$ and $E_2$.

Before turning to more complicated examples, note that Kirchoff 's rules have a direct analogy in closed loop, liquid flow systems. The voltage source becomes a pump, and the nodes become pipe fittings. Pressure is the analog of voltage and liquid flow rate is the analog of current. Grounding an electrical circuit is like opening a portion of the piping system to atmospheric pressure. Kirchoff 's first rule is equivalent to saying that the flow into an elbow, tee, or cross equals the flow out of it. Kirchoff 's second rule states that the net pressure drop around a closed loop is zero. Equivalently, the pressure at a point is independent of the path that the fluid used to reach that point.

**EXAMPLE 6.4** Refer to the previous figure and suppose $E_1$ is a resistor, $E_2$ is a capacitor, and $v_g(t) = v_o U(t)$. Find i(t). The system is initially relaxed.

**Solution:** Applying Kirchoff 's first rule at the upper node gives

$$i = i_R + i_C \qquad \text{(i)}$$

where we have set $i_1 = i_R$, $i2 = i_C$. The first rule is automatically satisfied at the lower node. Applying Kirchoff 's second rule gives

$$v_C = v_R = v_g(t) = v_o U(t)$$

or

$$\frac{1}{C}\int_0^t i_C dt = i_R R = v_0 U(t) \qquad \text{(ii)}$$

The three unknowns are i, $i_R$, and $i_C$. The three equations are Equation i and the two independent members of Equation ii. Since i(t) is desired, we will eliminate $i_R$ using

$$i_R = \frac{v_o U(t)}{R}$$

and will eliminate $i_C$ using

$$i_C = \frac{d}{dt}\int_0^t i dt = Cv_o \frac{dU(t)}{dt} = Cv_o \delta(t)$$

substituting into Equation i gives

$$i(t) = Cv_o \delta(t) + \frac{v_o U(t)}{R}$$

This curious looking result is in fact correct given the ground rules of the mathematical game we are playing. The voltage source is assumed to hold $v_o$ regardless of the current. The flow through the resistor comes to its steady state value immediately, and an instantaneous surge of current charges the capacitor immediately.

---

**EXAMPLE 6.5** Determine the current through the inductor given a voltage input of $v_o U(t)$.

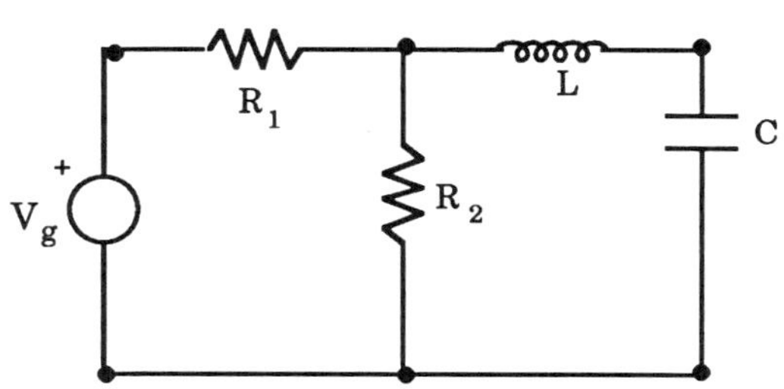

**Solution:** We first label the currents in a manner that automatically satisfies Kirchoff 's first rule:

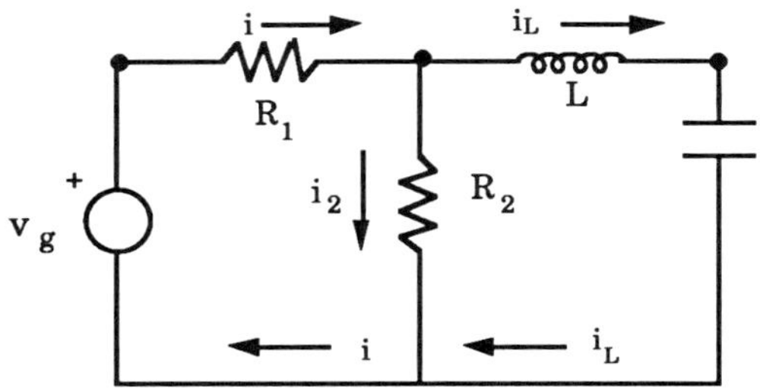


where $i = i_2 + i_L$.

We now apply Kirchoff 's second rule to the loops. There are three loops that can be used.

The inner loop containing the resistors and the voltage source gives

$$v_g = v_{R_1} + v_{R_2} \tag{i}$$

The outer loop containing no voltage source gives

$$v_{R_2} = v_L + v_C \tag{ii}$$

The overall loop gives

$$v_g = v_{R_1} + v_L + v_C \tag{iii}$$

Note that only two of these equations are independent. We use Equations ii and iii to give

$$i_2 R_2 = L\frac{di_L}{dt} + \frac{1}{C}\int_0^t i_L \, dt$$

and

$$v_o U(t) = iR_1 + L\frac{di_L}{dt} + \frac{1}{C}\int_0^t i_L dt$$

In the Laplace domain these become

$$I_2R_2 = LsI_L + \frac{I_L}{Cs}$$

$$\frac{v_o}{s} = IR_1 + LsI_L + \frac{I_L}{Cs}$$

We also have

$$I = I_2 + I_L$$

There are three equations in the three unknowns, I, $I_L$, and $I_2$. $I_L$ is desired. Eliminating I and $I_2$ gives

$$I_L = \frac{CR_2v_o}{(R_1 + R_2)\,CLs^2 + R_1R_2\,Cs + (R_1 + R_2)}$$

$$= \frac{(v_o/L)\left(\dfrac{R_2}{R_1+R_2}\right)}{s^2 + \left(\dfrac{R_1R_2}{R_1+R_2}\right)\dfrac{s}{L} + \dfrac{1}{CL}}$$

Inversion is left to the reader. Compare this result to that in Example 6.3 by taking the limit as $R_2 \to \infty$.

---

**EXAMPLE 6.6** Determine the steady state voltage drop across resistor R in the following circuit.

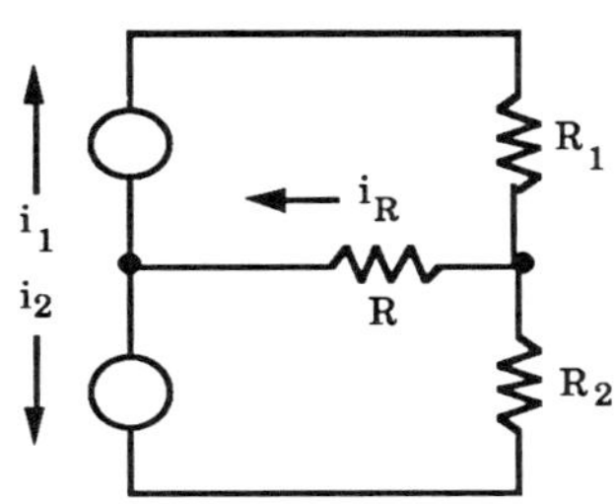


**Solution:** This problem involves two sources of constant current. Kirchoff 's first rule gives

$$i_R = i_1 + i_2$$

The desired voltage drop is

$$v_R = i_R R$$

$$= (i_1 + i_2) R$$

---

**EXAMPLE 6.7** Determine the steady state current flow through resistor R in the following circuit.

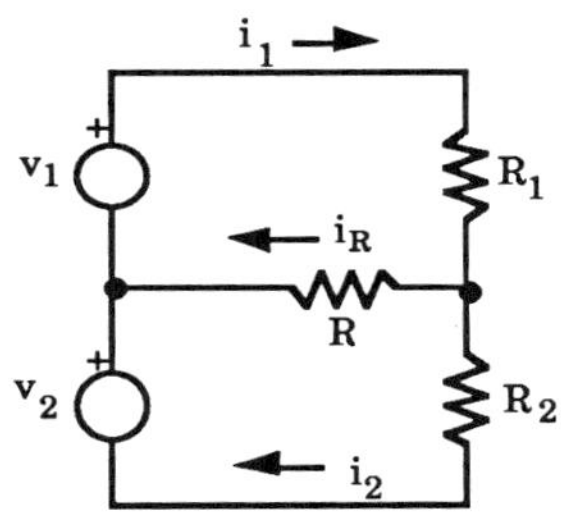


**Solution:** This problem involves two constant voltage sources in series. Kirchoff 's first rule gives

$$i_2 + i_R = i_1$$

Kirchoff 's second rule may be applied to any two of the three loops. The upper loop gives

$$v_1 = i_1 R_1 + i_R R$$

The lower loop gives

$$v_2 = i_2 R_2 - i_R R$$

These equations add to give

$$v_1 + v_2 = i_1 R_1 + i_2 R_2$$

which could have been obtained directly by applying Kirchoff 's second rule to the outside loop. The unknowns are $i_1$, $i_2$, and $i_R$. We eliminate $i_1$ and $i_2$ to obtain

$$i_R = \frac{R_2 v_1 - R_1 v_2}{R_1 R_2 + R R_1 + R R_2}$$

---

Examples 6.4 through 6.7 illustrate the use of Kirchoff 's rules in a variety of transient and steady state situations. The general approach is the same for all cases. The first rule is applied at each node. The second rule is applied for each loop. This procedure will generate enough **independent** equations to determine the unknown quantities by simultaneous solution. The simultaneous equations will be linear, algebraic equations in the Laplace domain. They will be linear, differential-integral equations in the time domain although they degenerate to algebraic equations in steady state situations such as Examples 6.6 and 6.7.

## 6.2 ALTERNATING CURRENTS

Although the methodology has been general, the examples considered so far have been for direct current. We now consider some alternating current cases where the input voltage has the form $v_0 \sin \omega t$ for $t > 0$. As before, we assume the system to be initially relaxed.

**EXAMPLE 6.8** Apply an alternating current to the series RCL circuit of Example 6.3.

**Solution:** Only the forcing function is changed. Instead of $dv_g/dt = v_o\delta(t)$ we now have $dv_g/dt = \omega \cos \omega t$. The governing ODE becomes

$$L\frac{d^2i}{dt^2} + R\frac{di}{dt} + \frac{i}{C} = v_o\, \omega \cos \omega t$$

and in the Laplace domain

$$Ls^2I + RsI + \frac{I}{C} = \frac{v_o\omega s}{s^2 + \omega^2}$$

or

$$I = \frac{(v_o/L)\, \omega s}{\left(s^2 + Rs/L + \dfrac{1}{CL}\right)(s^2 + \omega^2)}$$

This is a fourth order system, the inversion of which depends on the roots of the transient factor $(s^2 + Rs/L + 1/CL)$. The transient factor will invert to a possibly complex exponential that contains a damping term, $\exp(-Rt/2L)$. The damping term ensures that the transient solution will eventually decay. In Example 6.3, there were no other terms. Thus, the DC current in a series RCL circuit eventually declines to zero.

In an AC circuit, we have the factor $(s^2 + \omega^2)$, which has purely imaginary roots. Upon inversion, this gives rise to a persistent oscillation. Inversion can be accomplished in the usual manner:

$$I = \frac{K_1}{s+\alpha} + \frac{K_2}{s+\beta} + \frac{K_3 + K_4 s}{s^2 + \omega^2}$$

From this we obtain

$$K_1(s+\beta)(s^2+\omega^2) + K_2(s+\alpha)(s^2+\omega^2) + K_3(s+\alpha)(s+\beta) + K_4 s(s+\alpha)(s+\beta) = \frac{v_o \omega s}{L}$$

Equating the various powers of s gives

$$K_1 + K_2 + K_4 = 0$$

$$K_1\beta + K_2\alpha + K_3 + K_4(\alpha+\beta) = 0$$

$$K_1\omega^2 + K_2\omega^2 + K_3(\alpha+\beta) + K_4\alpha\beta = \frac{v_o\omega}{L}$$

$$K_1\beta\omega^2 + K_2\alpha\omega^2 + K_3\alpha\beta = 0$$

The determination of the coefficients is straightforward but tedious. The transient portion of the solution will be overdamped, critically damped, or underdamped depending on whether $R>2\sqrt{L/C}$, $R = 2\sqrt{L/C}$, or $R< 2\sqrt{L/C}$. See Example 6.3. The remaining portion of the solution will be a persistent oscillation:

$$i_\infty(t) = \frac{v_o}{\sqrt{R^2 + \left(\omega L - \dfrac{1}{\omega C}\right)^2}} \sin(\omega t - \phi)$$

where

$$\phi = \tan^{-1} \omega\left(\frac{L}{R} - \frac{1}{RC}\right)$$

The quantity

$$Z = \sqrt{R^2 + \left(\omega L - \frac{1}{\omega C}\right)^2}$$

is the **impedance**. Note that the impedance is frequency dependent and will be a minimum when

$$\omega L - \frac{1}{\omega C} = 0$$

or

$$\omega = \frac{1}{\sqrt{LC}}$$

This result gives the **resonant frequency** of a series RCL circuit. Note that $\phi = 0$ for this case so that the AC current is **in phase** with the applied voltage, and that $Z = R$ so that the impedance is equal to the DC resistance. The impedance increases as the frequency of the applied voltage deviates from the resonant frequency. The range of frequencies for which $Z < \sqrt{2}\, Z_{resonant}$ is called the **bandwidth** of the circuit.

Throughout this book we express ω as an **angular frequency** with units of [radians/time]. The ordinary frequency has units of [cycles/time] and is related to ω by

$$\text{Frequency} = \frac{\omega}{2\pi}$$

---

## PROBLEMS

6.1 Determine the current, i(t), for the following initially relaxed circuit subject to an input voltage of $v_g = v_o U(t)$:

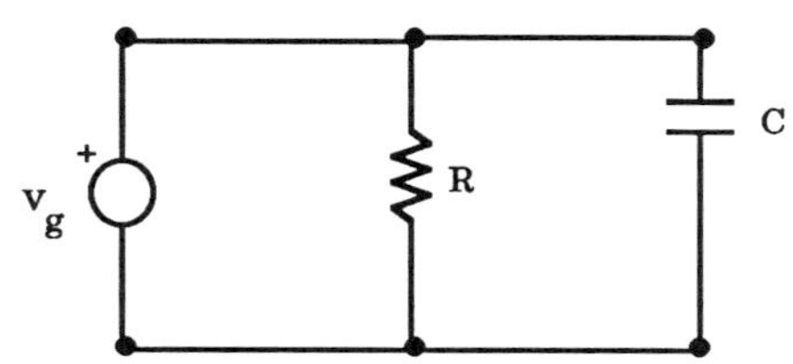

6.2 Find $I(s) = \pounds[i(t)]$ for the following initially relaxed circuit subject to $v_g = v_o U(t)$.

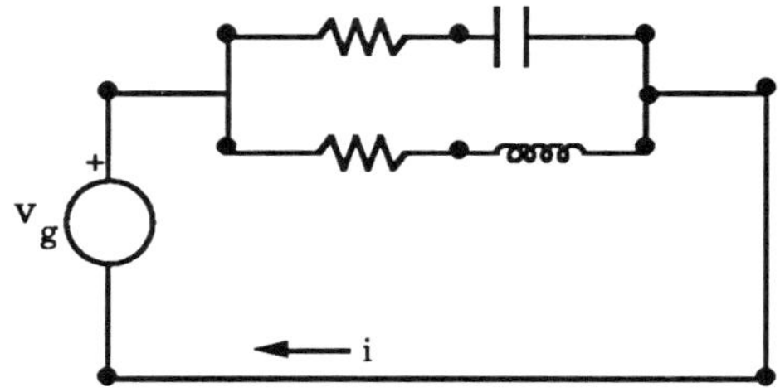


Note that the two resistances are equal.

6.3 Repeat Example 6.2 for an AC voltage, $v(t) = v_o \sin \omega t$. What is the impedance for the circuit?

6.4 Determine $i_L(t)$ for the following initially relaxed circuit given that $v_1(t) = v_2(t) = v_0 U(t)$.

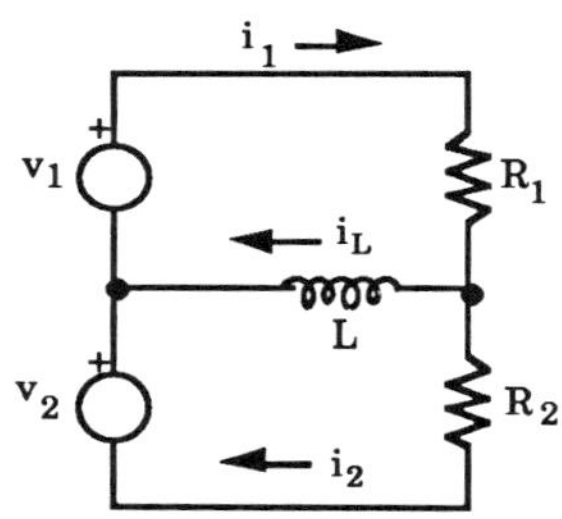


6.5 The RCL circuit in Example 6.8 is a **filter**. To understand this, pick values for R, C, and L and then plot $i(\infty)$ versus $\omega$.

# CHAPTER 7

# HEAT AND MASS TRANSFER

Chemical engineering treats a subject called **transport phenomena** which unifies the transfer of heat, mass, and momentum within a common conceptual framework. We now consider heat and mass transfer in lumped parameter systems. The topics of heat and mass transfer are almost completely analogous. They are governed by the same equations; only the notation is different. Our purpose here is not to treat these topics in detail but merely to show how ODEs arise in simple, dynamic problems involving heat and mass transfer.

## 7.1 CONVECTIVE HEAT TRANSFER

Suppose a copper ball of temperature $T_b$ is lowered into a stirred tank containing fluid of temperature $T_f$. How fast does the temperature of the ball change?

We have chosen the ball to be copper – an excellent heat conductor – so that its temperature can be considered uniform. The ball is lowered (quickly) into a liquid which is well mixed and thus also has a uniform temperature. Heat flow between the ball and the fluid is governed by **Newton's Law of Cooling**.

$$q = H\,(T_b - T_f) \qquad (7.1)$$

where q is positive when heat is flowing from the ball to the fluid, $T_b > T_f$. Here, H is a constant. It has units of [heat]/[time]/[temperature]. Often, we will find that H is proportional to the contact area between the ball and the fluid,

$$q = hA_b\,(T_b - T_f) \qquad (7.2)$$

where h is known as the **heat transfer coefficient** and has units of [heat]/[area]/[time]/[temperature]. SI units for h are J $m^{-2}$ $s^{-1}$ $K^{-1}$. Traditional units are Btu/$ft^2$/hr/°F.

Newton's law of cooling ignores the details of the temperature gradients that must exist in the fluid near the ball. More detailed consideration of this interface can lead to theoretical expressions for h. For now we regard h as an empirical constant and proceed to study how $T_b$ and $T_f$ vary with time. To do this we need enthalpy relationships for the various (lumped) masses in the system. Assume only sensible heat changes, no phase changes. The energy balance becomes

$$V\rho c \frac{dT}{dt} = q_{net} \tag{7.3}$$

where V, ρ, and c are the volume, density, and heat capacity for the mass under consideration; T is its temperature; and $q_{net}$ is the net, algebraic flow of heat [energy/time] into the mass.

We will use subscripts to distinguish between the various masses (or subsystems) in a system.

**EXAMPLE 7.1** Suppose the thermal capacity of the ball is small compared to that of the fluid, $V_b \rho_b c_b << V_f \rho_f c_f$, so that $T_f$ is approximately constant. Find $T_b(t)$.

**Solution:** Equation 7.3 gives an enthalpy relationship for the ball of radius R:

$$\frac{4}{3}\pi R^3 \rho_b c_b \frac{dT_b}{dt} = q_b$$

There is only one source of heat flow into the ball. This heat comes from the fluid. Its rate of flow is governed by Equation 7.2.

$$q_b = h(4\pi R^2)(T_f - T_b)$$

Eliminating $q_b$ gives

$$\frac{dT_b}{dt} + \left(\frac{3h}{R\rho_b c_b}\right) T_b = +\left(\frac{3h}{R\rho_b c_b}\right) T_f$$

or

$$\frac{dT_b}{dt} + \alpha T_b = +\alpha T_f$$

The solution to this first order ODE, subject to the initial condition that $T_b = T_b(0)$ at $t = 0$, is

$$\frac{T_b - T_f}{T_b(0) - T_f} = \exp(-\alpha t) = \exp\left(-\frac{3ht}{R\rho_b c_b}\right)$$

---

The above example treats the simplest possible case: a small, high conductivity ball loses heat in a well mixed sea of fluid. The heat loss is a first order process with time constant $R\rho_b c_b/3h$. (The reader should verify that $\alpha$ has units of $[\text{time}]^{-1}$.) The problem becomes somewhat more complicated if the volume of the fluid is finite, if there is a net flow of fluid through the stirred tank, or if the tank is jacketed. We consider a few cases as examples.

**EXAMPLE 7.2** Suppose the tank is insulated and contains volume V of a fluid of density $\rho_f$ and heat capacity $c_f$. Find $T_b(t)$.

**Solution:** In this case both $T_b$ and $T_f$ vary with time. The heat balance on the ball is unchanged.

$$\frac{dT_b}{dt} + \alpha T_b = +\alpha T_f \qquad \text{(i)}$$

We also need a heat balance on the fluid. The heat gained by this fluid equals the heat lost by the ball. Thus,

$$q_f = V_f \rho_f c_f \frac{dT_f}{dt} = -q_b = -(4\pi R^3 \rho_b c_b/3)\frac{dT_b}{dt} \qquad \text{(ii)}$$

Equations i and ii form a pair of simultaneous equations. Their solution can be accomplished in either the time domain or the Laplace domain. Notation in the Laplace domain is a problem since we are already using an upper case letter, $T_b$, for the dependent variable in the time domain. This fact points out an unpleasant truth: it is impossible to have a notation which is simultaneously simple, natural, and consistent.

For the problem at hand, we avoid notation problems by staying in the time domain. Equations i and ii contain two unknowns: $T_b$ and $T_f$. The usual method of solving ODEs is to differentiate to obtain a higher order of ODE in a single unknown. Differentiate Equation i and substitute Equation ii to obtain

$$\frac{d^2T_b}{dt^2} + \alpha\frac{dT_b}{dt} = \alpha\frac{dT_f}{dt} = -\alpha\beta\frac{dT_b}{dt}$$

or

$$\frac{d^2T_b}{dt^2} + (\alpha + \alpha\beta)\frac{dT_b}{dt} = 0 \tag{iii}$$

where

$$\beta = \frac{4\pi R^2 \rho_b c_b}{3V_f \rho_f c_f} \tag{iv}$$

Now, Equation iii can be integrated, but a final solution requires knowledge of $dT_b/dt$ at $t = 0$. This initial value is not known directly. It can be found by substituting back into Equation ii, integrating, and then matching solutions. The interested reader may wish to try this approach. Laplace transformation of Equations i and ii avoids the difficulty but introduces the previously mentioned notation problem. There are many ways to skin a cat. Time can be eliminated from Equation ii to give

$$\frac{dT_f}{dT_b} = -\beta$$

where $T_f = T_f(0)$ when $T_b = T_b(0)$. This integrates to

$$T_f = T_f(0) + \beta T_b(0) - \beta T_b \tag{iv}$$

The result is substituted into Equation i to give

$$\frac{dTb}{dt} + (\alpha + \alpha\beta)T_b = \alpha T_f(0) + \alpha\beta T_b(0)$$

This single ODE has a known initial condition: $T_b = T_b(0)$ at $t = 0$. The solution is

$$\frac{(1+\beta)T_b(t) - T_f(0) - \beta T_b(0)}{T_b(0) - T_f(0)} = e^{-(\alpha+\alpha\beta)t} \tag{v}$$

---

**EXAMPLE 7.3** The result of Example 7.2, although presumably correct, is complex and rather unpleasant in appearance. How can it be verified?

**Solution:** Formal verification requires back substitution into the governing ODEs. One must also show that the initial conditions are satisfied. This should be done; but before investing the effort, a faster plausibility check can be found by taking various limits.

Suppose $\alpha \to 0$ which would be the case of a very large ball in a small tank. Then Equation v reduces to $T_b(t) = T_b(0)$, which is obviously true for this limiting case. Suppose $\beta \to 0$, which would be the case of a small ball in a very large tank. Then the limit is the same as Example 7.1. Suppose $t \to \infty$. Then $T_b = [T_f(0) + \beta T_b(0)]/(1 + \beta)$. Is this correct? It is a steady value that can be calculated without appeal to the dynamic equations.

---

**EXAMPLE 7.4** Suppose the stirred tank in Example 7.2 is jacketed. The jacket temperature is held at a constant value of $T_j$. Heat transfer from the jacket to the fluid within the tank is governed by $q = U_jA_j(T_j - T_f)$, where $U_j$ is the heat transfer coefficient between the jacket and the fluid and $A_j$ is the contact area between the jacket and the fluid. Find the temperature of the ball as a function of time.

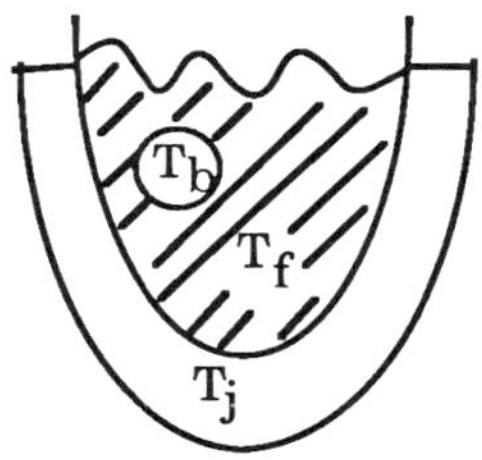


**Solution:** The governing equation for the temperature of the ball is unchanged.

$$\frac{dT_b}{dt} + \alpha T_b = \alpha T_f \tag{i}$$

The fluid loses heat to the ball but gains heat from the jacket. Thus,

$$V_f \rho_f c_f \frac{dT_f}{dt} = q_{net} = q_j - q_b$$

$$= U_jA_j(T_j - T_f) - (4\pi R^3 \rho_b c_b/3)\frac{dT_b}{dt} \tag{ii}$$

In deriving this result, we have assumed directions for the heat flows. If these assumptions are wrong, Equation ii will remain correct, but the q will change sign. The treatment of heat is thus similar to the treatment of current in Chapter 6.

Proceeding from this point requires the simultaneous solution of Equations i and ii. Although this certainly can be done in the time domain, it is enough easier in the Laplace domain to justify a change in our usual notation. Let

$$\tau_b = \tau_b(s) = \pounds[T_b(t)]$$

$$\tau_f = \tau_f(s) = \pounds[T_f(t)]$$

Then Equations i and ii become

$$s\tau_b - T_b(0) + \alpha\tau_b = \alpha\tau_f$$

$$V_f\rho_f c_f s\tau_f - V_f\rho_f c_f T_f(0)$$

$$= U_j A_j\left(\frac{T_j}{s} - \tau_f\right) - (4\pi R^3\rho_b c_b/3)\,[s\tau_b - T_b(0)]$$

or

$$s\tau_f - T_f(0) = \gamma\left[\frac{T_j}{s} - \tau_f\right] - \beta\,[s\tau_b - T_b(0)]$$

where

$$\gamma = \frac{U_j A_j}{V_f\rho_f c_f}$$

This set can be solved for $\tau_f$ and $\tau_b$. For $\tau_b$,

$$\tau_b = \frac{s^2 T_b(0) + s[(\gamma+\alpha\beta)T_b(0) + \alpha T_f(0)] + \alpha\gamma T_j}{s\,[s^2 + (\gamma+\alpha+\alpha\beta)s + \alpha\gamma]} \qquad \text{(iii)}$$

Inversion of this result is straightforward but messy. Before investing the necessary effort, the reader may want to confirm that Equation iii behaves properly in various limits. Try $\gamma \rightarrow 0$ and $t \rightarrow \infty$. Use the final value theorem in Table 4.2.

---

## 7.2 CONVECTIVE MASS TRANSFER

Consider a stirred tank with an oxygen sparger.

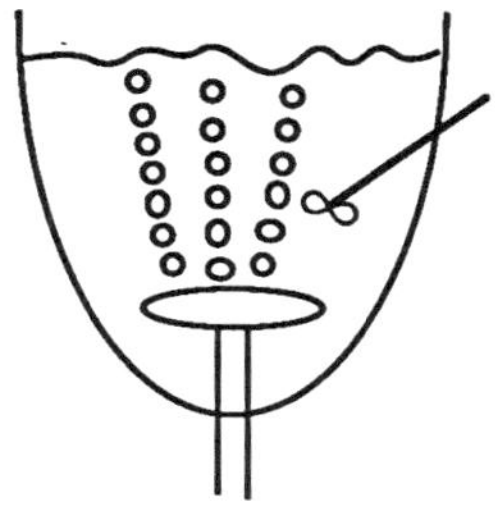

Oxygen bubbles through the liquid in the tank and gradually saturates it. What governs the rate of saturation? Analogous to Equation 7.2 we have

$$q_m = k_l A_b (a^* - a_f) \tag{7.4}$$

where $q_m$ represents the mass transfer rate, [moles/time], and $k_l$ is a mass transfer coefficient, [moles/time/area/concentration driving force]. $A_b$ is the total area for all the bubbles, $a_f$ is the concentration of oxygen in the liquid phase, and $a^*$ is the saturation concentration.

It is common practice to rewrite Equation 7.4 as

$$q_m = k_l A_i V (a^* - a_f) \tag{7.5}$$

where $A_i$ is the interfacial area (between bubbles and liquid) **per unit volume of tank**. Then $k_l A_i$ has units of $[\text{time}]^{-1}$. It is this composite quantity that is usually measured experimentally and for which correlations exist. In what follows, note that $k_l A_b$ can be replaced by $k_l A_i V$ without changing the results.

Analogous to Equation 7.3, the mass balance is

$$V_f \frac{da_f}{dt} = (q_m)_{net} \tag{7.6}$$

For the case now being considered, $q_m = (q_m)_{net}$ since oxygen transfer occurs only from the bubble phase to the liquid phase. More generally, there could be a third phase, or oxygen could be consumed by reaction. Also, there could be a net flow of liquid through the stirred tank. A variety of cases will be considered in the following examples. The reader should not be misled into thinking these examples are comprehensive. For mass transfer, as for heat transfer, a great many practical systems can be described by the techniques of this book. Our stress is on these techniques and not on specific applications.

**EXAMPLE 7.5** A tank of initially deaerated water is sparged with oxygen. Describe how the oxygen concentration in the liquid varies as a function of time.

**Solution:** Here $q_m = (q_m)_{net}$ so that Equations 7.4 and 7.6 combine to give

$$V_f \frac{da_f}{dt} = k_l A_b \,(a^* - a_f) \qquad \text{(i)}$$

Both $A_b$ and $a^*$ are assumed constant. $A_b$ is constant because the bubble population in the stirred tank is assumed to be at steady state. $a^*$ is constant because it is an equilibrium property; it is the concentration of oxygen in the water which would be in equilibrium with a gas phase of pure oxygen. Note that we are ignoring depletion of oxygen within bubbles and pressure effects due to the height of the liquid. Our results will not apply to a very deep system.

Equation i can be written as

$$\frac{da_f}{dt} + \frac{k_l\, A_b}{V_f} a_f = \frac{k_l\, A_b}{V_f} a^*$$

or

$$\frac{da_f}{dt} + \alpha a_f = \alpha a^*$$

The initial condition is $a_f = 0$ at $t = 0$ so that the system is initially relaxed. Laplace transformation gives

$$sA_f + \alpha A_f = \frac{\alpha a^*}{s}$$

or

$$A_f = \frac{\alpha a^*}{s(s+\alpha)}$$

The inverse is

$$a_f = a^*[U(t) - e^{-\alpha t}$$

$$= a^*\left[1 - \exp\left(-\frac{k_l A_B t}{V}\right)\right]$$

Thus, the sparged vessel approaches steady state in the usual manner for a first order system. This form of approach is sometimes called a **first order time lag.**

---

**EXAMPLE 7.6** Repeat Example 7.5 assuming now that oxygen is consumed in the liquid phase by a first order reaction with rate $ka_f$.

**Solution:** The oxygen transfer rate now contributes to both saturation and reaction.

$$(q_m)_{net} = \frac{Vda_f}{dt} + Vka_f = k_l A_b (a^* - a_f)$$

As before, $A_b$ and $a^*$ are assumed constant and the system is assumed to be initially relaxed with $a_f = 0$ at $t = 0$. The governing ODE is

$$\frac{da_f}{dt} + \left(k + \frac{k_l A_b}{V_b}\right) a_f = \frac{k_l A_b}{V_b} a^*$$

or

$$\frac{da_f}{dt} + (k + \alpha) a_f = \alpha a^*$$

The solution is

$$a_f = \frac{\alpha a^*}{k + \alpha} [1 - e^{-(k+\alpha)t}]$$

where $\alpha = k_l A_b / V_b$

---

**EXAMPLE 7.7** Repeat Example 7.5 with a net throughput, $Q_{in} = Q_{out} = Q$, of liquid. No reaction occurs and assume no entrainment of bubbles in the outlet stream. The dissolved oxygen concentration in the entering liquid is $a_{in}$ and this is also the initial concentration in the resident liquid. Find $a_{out}(t)$.

**Solution:** We seek the dynamic response of a stirred tank oxygenator. The net transfer rate contributes to both the resident fluid and the exiting fluid

$$(q_m)_{net} = V \frac{da_f}{dt} + Qa_f$$

where we have set $a_{out} = a_f$ since the tank is well stirred. There are now two sources of oxygen, transfer from the bubbles, and dissolved oxygen in the inlet flow.

$$k_l A_b (a^* - a_f) + Q a_{in} = (q_m)_{net}$$

Combining these equations gives the governing ODE

$$V \frac{da_f}{dt} + (Q + k_l A_b) a_f = k_l A_b a^* + Q a_{in}$$

or

$$\frac{da_f}{dt} + \left( \frac{1}{\bar{t}} + \alpha \right) a_f = \frac{a_{in}}{\bar{t}} + \alpha a^*$$

The initial condition is $a = a_{in}$ at $t = 0$. The Laplace transform is

$$sA_f - a_{in} + \left( \frac{1}{\bar{t}} + \alpha \right) A_f = \frac{a_{in}/\bar{t} + \alpha a^*}{s}$$

or

$$A_f = \frac{a_{in}/\bar{t} + \alpha a^*}{s(s + \alpha + 1/\bar{t})} + \frac{a_{in}}{s + \alpha + 1/\bar{t}}$$

The inverse gives the total response of the oxygenator.

$$a_f = a_{out}(t) = \frac{a_{in} + \alpha \bar{t} a^* + \alpha \bar{t} (a_{in} - a^*) e^{-(1+\alpha \bar{t}) t/\bar{t}}}{1 + \alpha \bar{t}}$$

The steady state response is

$$a_{out}(\infty) = \frac{a_{in} + \alpha \bar{t} a^*}{1 + \alpha \bar{t}}$$

---

## 7.3 STEADY STATE CONDUCTION AND DIFFUSION

The governing equation for unsteady heat transfer in a three-dimensional solid is

$$\frac{\partial T}{\partial t} = D_T \left( \frac{\partial^2 T}{\partial x^2} + \frac{\partial^2 T}{\partial y^2} + \frac{\partial^2 T}{\partial z^2} \right) \tag{7.7}$$

where $D_T$ is the thermal diffusivity. The governing equation for unsteady diffusion in a three-dimensional solid is

$$\frac{\partial a}{\partial t} = D_m \left( \frac{\partial^2 a}{\partial x^2} + \frac{\partial^2 a}{\partial y^2} + \frac{\partial^2 a}{\partial z^2} \right) \tag{7.8}$$

where $D_m$ is the molecular diffusivity.

Most applications of these equations are beyond the scope of this text. We did see in Chapter 5 how PDEs could be reduced to ODEs through use of the Laplace transform. Here, we consider the situation where a PDE becomes an ODE due to the inherent simplicity of the problem. This situation arises in **steady state diffusion** (or conduction) in **one spatial dimension.**

In one spatial dimension, say x, terms like $\partial^2 a/\partial y^2$ and $\partial^2 a/\partial z^2$ vanish; and at steady state, $\partial a/\partial t = 0$. Thus, the diffusion equation reduces to

$$\frac{d^2 a}{dx^2} = 0$$

This equation admits the following general solution

$$a = c_1 x + c_2$$

where $c_1$ and $c_2$ are constants to be determined from the boundary conditions. About the only problem of physical significance governed by this result is diffusion across a stagnant film or through a membrane. Suppose $a = a_o$ at $x = 0$ and $a = a_L$ at $x = L$. Then

$$a = \frac{(a_L - a_o)\, x}{L} + a_o \tag{7.9}$$

so that the concentration profile across a membrane is a linear function of distance. The rate of mass transfer, or flux, is given by

$$J_m = - D_m \frac{da}{dx} = -D_m \frac{(a_L - a_o)}{L} \tag{7.10}$$

where $J_m$ has units of [moles/area/time]. The minus sign in Equation 7.10 shows that diffusion occurs in the direction opposite to a concentration gradient. Thus, if $a_L - a_o > 0$, then $da/dx > 0$ and diffusion will occur in the negative x direction.

Equations similar to 7.9 and 7.10 govern the conduction of heat across an insulating slap. However, the flux equation is subtly different:

$$J_h = -\kappa \frac{dT}{dx} = -\kappa \frac{(T_L - T_o)}{L} \tag{7.11}$$

where $\kappa$ is the thermal conductivity with units of [energy/length/time/temperature driving force]. Equation 7.11 contains the thermal conductivity, $\kappa$, rather than the thermal diffusivity, $D_T = \kappa/\rho C$. Equation 7.10 contains the molecular diffusivity, $D_m$. The reason for the difference goes back to Equation 7.7. Molecules are diffusing and a is the concentration of molecules per unit volume. In Equation 7.8, heat is diffusing but T is the temperature and not (directly at least) the concentration of heat.

## PROBLEMS

7.1 Suppose the insulated stirred tank in Example 7.2 has inlet and outlet flows, $Q_{in} = Q_{out} = Q$, with inlet temperature $T_{in} = T_f(0)$. Find $T_b(t)$.

7.2 A membrane with width $L_1$ and diffusivity $D_1$ is adjacent to a membrane of width $L_2$ and diffusivity $D_2$. The concentration at one side of the combined membranes is $a_0$. The concentration at the other side is $a_L$. What is the magnitude of the flux?

7.3 What is the heat conduction analogue of Equation 7.9?

7.4 The copper ball in Example 7.1 is coated with a thin layer of thermal insulation having thickness $\Delta r$ and thermal conductivity $\kappa$. Revise Example 7.1 to account for the insulation. Use a pseudo steady state approximation for the temperature profile within the insulation.

7.5 Prove that the time domain solution to Example 7.4 will not oscillate.

7.6 Rework Example 7.2 for two copper balls of equal size but of possibly different initial temperatures, $T_1$ and $T_2$. Discuss the case of $T_1(0) - T_f(0) = T_f(0) - T_2(0)$.

# CHAPTER 8

# ELEMENTARY MECHANICAL SYSTEMS

This chapter derives the equations of motion for some simple mechanical systems. Like most of the application chapters in this book, our treatment will be cursory. The main goal is to analyze the dynamics of rigid bodies using differential equations and, where appropriate, Laplace transformation. We also analyze some simple flow systems.

## 8.1 NEWTON'S SECOND LAW

Arguably the most important result in the history of science is

$$f_{net} = \frac{d(mv)}{dt} \tag{8.1}$$

where $f_{net}$ is the net force acting on a body of mass, m, having velocity, v. We usually deal with systems of constant mass. Then Newton's second law takes the familiar form

$$f_{net} = ma = m\frac{dv}{dt} = m\frac{d^2x}{dt^2} \tag{8.2}$$

where a is the acceleration and x is the displacement of the body. In the Laplace domain

$$\begin{aligned} F_{net}(s) &= mA(s) = m[s(V(s) - v(0+)] \\ &= m[s^2X(s) - sx(0+) - v(0+)] \end{aligned} \tag{8.3}$$

Much of elementary mechanics is concerned with determining $f_{net}$ for systems of interest. The reader will recall the inclined planes and simple pendulums of introductory physics. If $f_{net}$ is constant, as say for a freely falling body, Equation 8.2 gives

$$a = f_{net}/m$$

$$v = v(0+) + \frac{f_{net}}{m}t \qquad (8.4)$$

$$x = x(0+) + v(0+)t + \frac{f_{net}\,t^2}{2\,m}$$

Here, of course, v and x increase without bound, and no steady or periodic state is achieved.

### 8.1.1 Frictional Resistance

Most bodies in motion encounter frictional resistance so that

$$f_{net} = f_a - f_{friction} \qquad (8.5)$$

where $f_a$ is the applied force, due say to gravity, and $f_{friction}$ is the frictional force, due say to air resistance, that acts in a direction to retard motion. In practical situations, functional expressions for $f_{friction}$ can be quite complex. They are generally nonlinear functions of velocity, v. Thus, as is the case with chemical reactions, most rigorous models of dynamic mechanical systems are nonlinear and must be linearized about some nominal velocity or else solved numerically.

A **linear frictional resistance** has the form

$$f_{friction} = Bv \qquad (8.6)$$

This approximation leads to linear equations of motion for which analytical solutions are possible. Mechanical engineers use Equation 8.6 as the model of a **dashpot**, which is an idealized kind of fluid dampening device or shock absorber.

**EXAMPLE 8.1** Solve the equation of motion for a system with constant mass, a constant applied force, and a linear frictional resistance.

**Solution:** The equation of motion becomes

$$f_{net} = f_a - Bv = m\frac{dv}{dt} \qquad (i)$$

The reader will recognize this as a first order system with time constant m/B. The Laplace domain solution is

$$v(s) = \frac{(f_a/m) + sv(0+)}{s(s+B/m)} \qquad \text{(ii)}$$

This system has a steady state value which can be found by setting dv/dt = 0 in Equation i or else from the final value theorem

$$v(\infty) = \lim_{s \to 0} sV(s) = \frac{f_a}{B}$$

The usual partial fraction expansion is

$$V(s) = \frac{K_1}{s} + \frac{K_2}{s + B/m} = \frac{(f_a/m) + sv(0+)}{s(s + B/m)}$$

or

$$K_1(s + B/m) + K_2 s = (f_a/m) + sv(0+)$$

Setting s = -B/m and s = 0 according to Equation 4.35 gives

$$K_2 = -f_a/B + v(0+), \qquad K_2 = f_a/B$$

The inverse is

$$v(t) = \frac{f_a}{B} U(t) + (-\frac{f_a}{B} + v_o e^{-Bt/m})$$

or

$$v(t) = \frac{f_a}{B} + (v_o - f_a/B)e^{-Bt/m}$$

The case $f_a = 0$ corresponds to a body initially in motion which is subject to linear frictional resistance. Then

$$v(t) = v_o e^{-Bt/m}$$

The body slows but never quite stops. Since real bodies stop, it may seem that a resistance linear in v is inappropriate for modeling sliding or rolling friction. See, however, Problem 8.1.

---

**EXAMPLE 8.2** Suppose a rocket, rising vertically, generates constant thrust $f_{thrust}$ and loses mass – due to fuel consumption – at a constant rate, k. The rocket is subject to gravity, g, and a linear friction resistance, Bv. Initial values are x = 0, v = 0, and m = $m_0$ at t = 0. Find the equation of motion for the rocket on its upward trajectory.

**Solution:** This is a variable mass problem and thus uses Newton's second law in the form

$$f_{net} = \frac{d(mv)}{dt}$$

where, by hypothesis m = $m_o$ - kt. The upward thrust is opposed by gravity and air resistance

$$f_{net} = f_{thrust} - mg - Bv$$

The equation of motion become

$$m\frac{dv}{dt} + v\frac{dm}{dt} = (m_o - kt)\frac{dv}{dt} - kv$$

$$= f_{net} = f_{thrust} - mg - Bv$$

or

$$(m_o - kt)\frac{dv}{dt} + (B - k)v = f_{thrust} - m_o g + kgt$$

This ODE is linear but has a variable coefficient. The reader is invited to solve it using the integrating factor method of Chapter 3.

---

### 8.1.2 Restoring Forces

We now consider forces which act to restore a body to some initial position. The simplest examples are springs or rubber bands. A linear model for their behavior is

$$f_{spring} = K(x-x_r) \tag{8.7}$$

where K is a constant known as the **stiffness**. Here $x_r$ is a coordinate position that corresponds to the relaxed spring with $f_{spring}$ = 0. Note that $f_{spring}$ will change sign as x crosses $x_r$.

For real springs or rubber bands to obey Equation 8.7, they must be in the linear or Hook's law region of their elastic behavior. Note also that real materials behave differently in tension than compression. To obtain a linear response in practice, the spring or rubber band must be under an initial load, $f_{equil}$, which gives an equilibrium displacement, $x_{equil}$. Thus Equation 8.7 is better written as

$$f_{restoring} = f_{spring} - f_{equil} = K(x - x_{equil}) \tag{8.8}$$

**EXAMPLE 8.3** A body of mass m is suspended by a massless spring. Find $f_{restoring}$ as a function of displacement.

**Solution:**

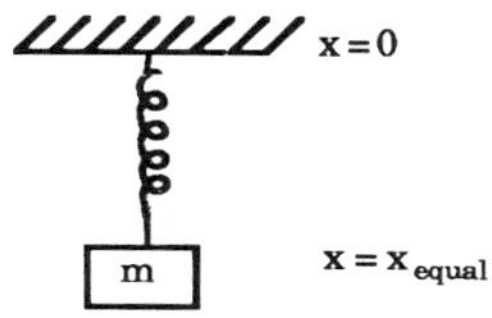


At equilibrium, the spring will be partially stretched so that the spring force just balances the force of gravity

$$f_{spring} - mg = 0$$

At this position, $x = x_{equil}$ and $f_{restoring} = 0$. If $x > x_{equil}$ then $f_{restoring} > 0$. The sign on $f_{restoring}$ is arbitrary. A common convention is for upward forces to be positive. Then for Equation 8.8 to predict the right sign (assuming $K > 0$), it is necessary to use a coordinate system which is positive in the downward direction. We are free to pick any fixed position for the point $x = 0$. One possible choice is shown in the above figure. Another choice is to set $x_{equil} = 0$ so that Equation 8.8 becomes

$$f_{restoring} = Kx$$

It is this version of Equation 8.8 that we will normally choose.

---

**EXAMPLE 8.4** Suppose that an additional downward force is applied to the suspended mass in Example 8.3., displacing it to a distance $x_0$ below the original equilibrium position at $x = 0$. At time $t = 0$, the additional force is suddenly released. Find x(t) given that the mass was initially at rest.

**Solution:** After the additional force is released, only the restoring force acts on the mass. Thus, Newton's second law becomes

$$m\frac{dv}{dt} = -Kx \tag{i}$$

(The reader may wonder what has happened to the force of gravity. We have build it into the definition of restoring force.) The negative sign arises because an upward force causes acceleration in the upward, negative x direction.

Equation i contains two dependent variables, v and x. To complete the model we need

$$\frac{dx}{dt} = v \tag{ii}$$

The initial conditions are $v = 0$ and $x = x_o$ at $t = 0$.

Equations i and ii must be solved simultaneously. Laplace transformation is as good a method as any:

$$msV = -KX$$

$$sX - x_o = V$$

Eliminating V between these gives

$$ms(sX - x_o) = -KX$$

or

$$X = \frac{msx_o}{ms^2 + K} = \frac{sx_o}{s^2 + K/m}$$

The inverse is

$$x = x_o \cos \omega t$$

where

$$\omega = \sqrt{K/m}\ .$$

A mass hung from a spring is the mechanical engineers' equivalent of an LC circuit. In the absence of air resistance or internal, frictional resistance, a mass suspended from a spring will oscillate forever. So will an LC circuit in the absence of electrical resistance. (Chemical engineering lacks such simple analogues of sustained oscillations.) Real circuits have resistance. So do real springs. Mechanical engineers model the internal resistance of a spring by placing a dashpot in parallel with the spring.

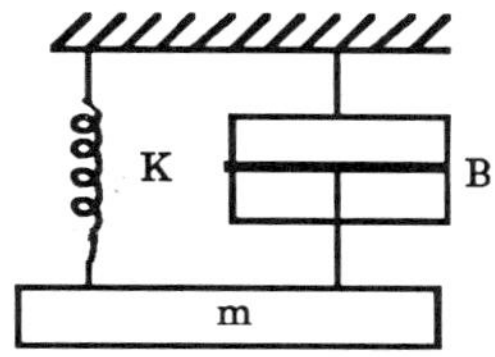


The dashpot imposes a frictional resistance which is linear with respect to velocity according to Equation 8.6.

---

**EXAMPLE 8.5** Model a mass suspended from a spring in parallel with a dashpot as indicated in the above figure. Suppose $x = x_o$ and $v = 0$ at $t = 0$.

**Solution:** As before, the force of gravity, -mg, is treated as an offset to the spring force. The net forces on the mass are

$$-Kx - Bv = m\frac{dv}{dt}$$

or

$$-KX - BV = msV$$

We also have

$$sX - x_o = V$$

Eliminating V gives

$$-KX - B(sX - x_o) = ms(sX - x_o)$$

or

$$X = \frac{msx_o + Bx_o}{ms^2 + Bs + K} = \frac{sx_o + (B/m)x_o}{s^2 + (B/m)s + K/m}$$

Factoring the denominator gives

$$X = \frac{sx_o + (B/m)x_o}{(s+s_1)(s+s_2)}$$

where

$$s_1, s_2 = \frac{-B \pm \sqrt{B^2 - 4K}}{2m}$$

Just as for the RCL circuit in Chapter 6, there are three cases to consider. If $B^2 - 4K < 0$, $s_1$ and $s_2$ will be complex conjugates; the system is said to be **underdamped**; x(t) will oscillate with frequency $(B^2 - 4K)/4m^2$ ; and the amplitude of the oscillation will decrease as exp(-B/2m). If $B^2 = 4K$, the system is **critically damped**; the roots are real and equal; and the equilibrium position, x = 0, will be approached without oscillation. If $B^2 > 4K$, the system is **overdamped**; the roots are real but unequal; no oscillation will occur; and the approach to equilibrium will be slower than for the critically damped case.

---

We turn now to multiple body problems where the system consists of several massive bodies that are coupled together by some arrangement of springs and dashpots. We will apply Newton's second law to the individual bodies, but will note that the coupling of the bodies means that the motion of one will cause a force on another. The basic approach is just a force balance on each body where any net, unbalanced force will lead to acceleration according to Equations 8.1 or 8.2.

**EXAMPLE 8.6** Write the equations of motion for three rail cars in series, assuming the cars are coupled by springs, that there is a rolling resistance proportional to velocity, and that the string of cars is subject to a constant motive force. A diagram for this situation is

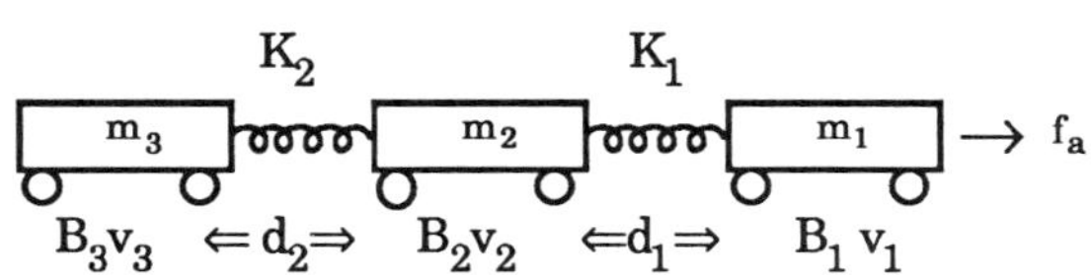


**Solution:** We apply a force balance to each of the masses. Car #1 is subject to a motive force $f_a$, a restoring force $K_1[d_1 - (d_1)_{equil}]$, and a frictional resistance of $B_1 v_1$.

$$m_1 \frac{d^2x_1}{dt^2} = f_a - K_1 [d_1 - (d_1)equil ] - B_1 \frac{dx1}{dt} \qquad \text{(i)}$$

where $x_1$ denotes the position of car #1. Car #2 is acted on by the same restoring force as for car #1 but the force is in the opposite direction. If $d_1 > (d_1)equil$, car #1 will be decelerated and car #2 will be accelerated. Car #2 is also subject to a restoring force due to the spring between it and car #3. Overall, for car #2 we have

$$m_2 \frac{d^2x_2}{dt^2} = K_1 [d_1 - (d_1)_{equil}] - K_2 [d_2 - (d_2)_{equil}] - B_2 \frac{dx_2}{dt} \qquad \text{(ii)}$$

and for car #3

$$m_3 \frac{d^2x_3}{dt^2} = K_2 [d_2 - (d_2)_{equil} ] - B_3 \frac{dx_3}{dt} \qquad \text{(iii)}$$

These three equations contain the car positions, $x_1$, $x_2$, and $x_3$, as dependent variables. The spring displacements can be related to the car positions by

$$d_1 = x_1 - x_2 \qquad \text{(iv)}$$

$$d_2 = x_2 - x_3 \qquad \text{(v)}$$

where we have ignored the lengths of the cars.

Equation i through v are a set of five equations in five unknowns, $x_1$, $x_2$, $x_3$, $d_1$, and $d_2$. This set contains two algebraic equations and three second order ODEs. The ODEs require six boundary conditions. These can be the three initial positions, $x_1$, $x_2$, and $x_3$, and three initial velocities, $dx_1/dt$, $dx_2/dt$, and $dx_3/dt$. Typical initial conditions would be zero velocities and zero restoring forces at $t = 0$. For zero restoring force

$$x_1 = (d_2)_{equil} + (d_1)_{equil}$$

$$x_2 = (d_2)_{equil}$$

$$x_3 = 0$$

With initial conditions known, Equations i through v can be solved simultaneously using Laplace transforms. The solution is algebraically complex. There is a pole at $s = 0$ plus six nontrival poles, which can be real or imaginary in many combinations.

---

A system consisting of N massive bodies gives rise to a set of N second order ODEs. This set is equivalent to a single ODE of order 2N. It is apparent that the dynamics of such a system can be very complex. This statement is true even when the various forces acting on the system are **colinear**. A simple example of a noncolinear system is

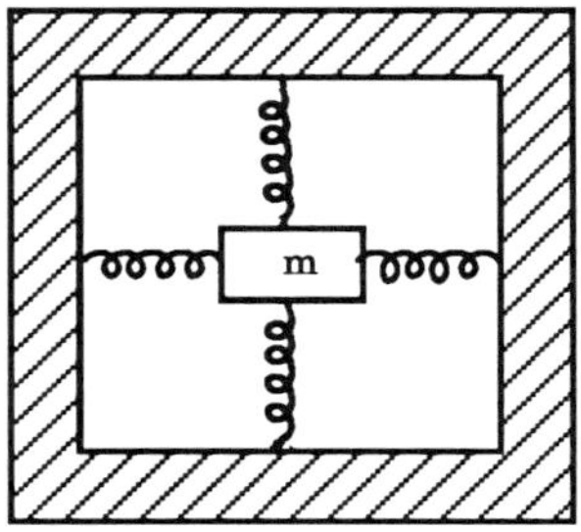


Such systems remind us that force, velocity, and acceleration are vectors. Newton's second law should really be written as

$$\mathbf{f}_{net} = \frac{d(m\mathbf{v})}{dt} \tag{8.9}$$

The component form of this for a constant mass system is

$$(f_x)_{net} = ma_x = m\frac{d^2x}{dt^2}$$

$$(f_y)_{net} = ma_y = m\frac{d^2y}{dt^2} \tag{8.10}$$

$$(f_z)_{net} = ma_z = m\frac{d^2z}{dt^2}$$

**EXAMPLE 8.7** Solve the two-dimensional equations of motion for the above illustrated situation.

**Solution:** We assume the spring constants to be equal and will ignore interactions between the spring forces in the x and y directions. The equations of motion are:

$$m\frac{d^2x}{dt^2} = -2Kx$$

$$m\frac{dy^2}{dt^2} = -2Ky$$

(Where does the factor of two come from?) Solutions are

$$x = x_o \sin\sqrt{\frac{2K}{m}}\,t$$

$$y = y_o \sin \sqrt{\frac{2K}{m}}\, t$$

The two motions are sinusoidal and independent.

---

## 8.2 CONSERVATION OF MECHANICAL ENERGY

To this point we have used force balances and Newton's second law to model simple mechanical systems. The principle of conservation of energy can also be used. In this chapter we will assume energy to have two strictly mechanical components: kinetic energy, $mv^2/2$, due to the velocity of a body, and potential energy, mgh, due to the presence of a gravitational field.

**EXAMPLE 8.8** Use conservation of energy to analyze the dynamics of a marble gently rolling off a table.

**Solution:** We will take the y-direction as positive upward and will ignore any motion in the x and z directions. Conservation of kinetic and potential energy gives

$$mv^2/2 + mgh = \text{constant}$$

The constant can be evaluated from the initial condition that v = 0 when the marble was at table height, $h_o$. Thus,

$$mv^2/2 + mgh = mgh_o \qquad \text{(i)}$$

Solving for v

$$v = \sqrt{2g(h_o - h)} = \frac{dh}{dt}$$

or

$$\frac{dh}{\sqrt{2g(h_o - h)}} = dt$$

Integrating and using the initial condition that $h = h_o$ at $t = 0$ gives

$$\sqrt{2g(h - h_o)} = gt$$

or

$$h = h_o - gt^2/2$$

which is the usual result for a freely falling object that was initially at rest.

---

The principles of energy and mass conservation are typically more useful in analyzing flow systems than in analyzing the motion of a single, isolated body. The energy balance for a flow system is

$$\text{Input} = \text{Output} + \text{Rate of Accumulation} \tag{8.11}$$

There is no energy generation or formation term in Equation 8.11 since we suppose this to be an overall balance encompassing all forms of energy. By comparison, a heat balance can have generation terms since other forms of energy are readily converted to heat.

The direct application of Equation 8.11 to a flow system requires a number of simplifying assumptions. We shall assume unidirectional flow of an incompressable, frictionless fluid. We begin with a steady state example.

**EXAMPLE 8.9** The water box on your toilet springs a pinhole leak at the bottom. Model the flow rate through the hole.

**Solution:** This prosaic example was chosen because a toilet has a level controller to maintain the liquid height at $h_o$. It is this that allows a steady state analysis.

The material balance is simply

$$Q_{in} = Q_{out} = Q.$$

The energy input is $\rho Q h_o g$ where we have ignored any kinetic energy of the inlet water. The energy output is all kinetic, $\rho Q v_{out}^2/2$. At steady state the accumulation is zero. Thus, Equation 8.11 gives

$$\rho Q h_o g = \rho Q v_{out}^2/2$$

or

$$v_{out} = \sqrt{2gh_o}$$

The flowrate is just $v_{out}A_{out}$ where $A_{out}$ is the area of the hole.

---

**EXAMPLE 8.10** Suppose the pinhole is enlarged so that its cross sectional area, $A_o$, is significant compared to the area of the water box, A. Also suppose the inlet supply of water is discontinued.

**Solution:** Now $h = h(t)$ so that an unsteady analysis with accumulation terms is needed. A mass balance gives

$$v_{out}A_{out} = vA \tag{i}$$

where v is the velocity of water in the box. Also,

$$v = \frac{dh}{dt} \tag{ii}$$

Note that v, $v_{out}$ and dh/dt will all be negative in this case and that

$$Q_{out} = -v_{out}A_{out} \tag{iii}$$

since we have supposed $Q_{out}$ to be positive.

The energy input is zero. The energy output is $\rho Q_{out}v_{out}^2/2$. The accumulation term is somewhat complicated as it has both potential and kinetic energy components. The average height of water is h/2 and the mass of water is $\rho hA$. Thus, the total potential energy is $\rho gh^2A/2$. The kinetic energy of water inside the box is $v^2/2$ per unit mass so that the total kinetic energy is $\rho hAv^2/2$. The energy balance becomes

$$0 = \rho Q_{out}v_{out}^2/2 + \frac{d}{dh}(\rho gh^2A/2 + \rho hAv^2/2)$$

Using Equations i – iii this eventually simplifies to

$$0 = -v_{out}^2/2\left(1 + (\frac{A_{out}}{A})^2\right) + 2gh + 2h\,(\frac{A_{out}}{A})\ v_{out}\ \frac{dv_{out}}{dt} \tag{iv}$$

If $A_{out}/A \rightarrow 0$, this result further simplifies to

$$v_{out} = \sqrt{2gh} \tag{v}$$

which is similar to the steady state case of Example 8.9. Equation v represents a form of pseudo-steady state where the outlet velocity immediately responds to changes in water level.

---

## PROBLEMS

8.1 A body initially at motion but subject to a linear friction resistance slows but never quite stops. See Example 8.1. Practical experience shows that rolling or sliding bodies do stop. Show how the actual stopping distance can be used to estimate a linear frictional coefficient, B.

8.2 Suppose that a body falling in a gravity field experiences a frictional force proportional to velocity, Equation 8.6. Derive an expression for its terminal velocity.

8.3 Cut one car off the train in Example 8.6. Suppose the remaining cars have equal mass and equal frictional resistance. Determine the intercar separation distance, d(t), for a start up from rest.

8.4 Introductory physics courses treat the two-dimensional trajectory of artillery shells by neglecting air resistance. Repeat this analysis assuming a friction law of the form $\mathbf{f}_{friction} = B\mathbf{v}$. Specifically, develop equations for x(t) and y(t) for a projectile launched with velocity $\mathbf{v}$ from location $x = 0,\ y = 0$.

8.5 A more sophisticated analysis of the four-spring problem in Example 8.7 suggests equations of the form

$$m \frac{d^2x}{dt^2} = -2K\,(x + \frac{xy}{L})$$

$$m \frac{d^2y}{dt^2} = -2K\,(y + \frac{xy}{L})$$

where L is the equilibrium length of the springs. Use computer simulation to compare these equations to those in Example 8.7.

8.6 What energy terms must be negligible for Equation v in Example 8.10 to hold rather than Equation iv? What is the initial condition on $v_{out}$ when the full Equation iv is used?

8.7 Find h(t) using the pseudo-steady state Equation v in Example 8.10.

# CHAPTER 9

# TRANSFER FUNCTIONS AND NETWORK ANALYSIS

This chapter describes a method for analyzing complex linear systems using Laplace transform techniques. The general method can be applied to all sorts of systems: electrical, mechanical, or chemical. However, the specifics are rather different since physical quantities such as voltage and concentration follow different rules of addition and subtraction. We will study two specific cases. One case is voltages in electrical circuits, the governing equations for which constitute the basis of **control theory**. The other case is concentrations in flow systems, the governing equations for which are basic to **mass balances** in chemical processes. A common aspect of both treatments is the concept of **transfer function**.

## 9.1 TRANSFER FUNCTIONS

We begin by defining a **system** as some physical device (or collection of devices) that takes an input, $a_{in}(t)$, and generates an output, $a_{out}(t)$. The input may be concentration, voltage, current, force, and so on. In Chapter 11 we will consider outputs that are different physical quantities than the input. For now, the output will be the same physical quantity as the input but a different function of time. We represent a system by the following diagram:

INPUT $\longrightarrow$ [ SYSTEM ] $\longrightarrow$ OUTPUT

$a_{in}(t)$ $\qquad\qquad$ $a_{out}(t)$

Mathematically, a system corresponds to some general operation that transforms $a_{in}(t)$ into $a_{out}(t)$. In what follows, we will restrict the system to operations where $a_{in}$ and $a_{out}$ are related by a linear ODE with constant coefficients. A general form for the time domain relationship between $a_{in}$ and $a_{out}$ is

$$\frac{d^n a_{out}}{dt^n} + C_{n-1}\frac{d^{n-1} a_{out}}{dt^{n-1}} + \dots + C_1 \frac{da_{out}}{dt} + C_o a_{out} = Ka_{in} \tag{9.1}$$

where K and the various Cs are constants. We can also write this as

$$\frac{d^n a_{out}}{dt^n} + (\alpha + \beta + \gamma + \dots)\frac{d^{n-1} a_{out}}{dt} + \dots + (\alpha\beta\gamma\dots)\, a_{out} = (\alpha\beta\gamma\dots)\, a_{in} \tag{9.2}$$

where $-\alpha$, $-\beta$, $-\gamma$, ... are the n roots of

$$s^n + C_{n-1}s^{n-1} + \dots + C_1 s + C_o = 0 \tag{9.3}$$

We now make an additional assumption that will be used throughout this chapter. The system is to be **initially relaxed** so that $a_{out}(t)$ and its first n-1 derivatives are zero at $t = 0$. A system which has had zero input for all negative time, $a_{in}(t) = 0$ for $t < 0$, will be initially relaxed.

We now take the Laplace transform of the governing ODE. Since the system is initially relaxed, this gives a particularly simple result:

$$\begin{aligned} A_{out}(s) &= \frac{KA_{in}(s)}{s^n + C_{n-1}s^{n-1} + \dots + C_1 s + C_o} \\ &= \frac{(\alpha\beta\gamma\dots)\, A_{in}(s)}{(s+\alpha)(s+\beta)(s+\gamma)\dots} \\ &= F(s)\, A_{in}(s) \end{aligned} \tag{9.4}$$

where F(s) is the **transfer function** for the system.

The important thing to note about Equation 9.4 is that the transfer function, F(s), does not depend on $A_{in}$. It is a property of the system and applies to any input function. Further,

$$\frac{A_{out}(s)}{A_{in}(s)} = F(s) \tag{9.5}$$

shows that, in the Laplace domain, the ratio of output to input is a property of the system, independent of $A_{in}(s)$.

To determine F(s), we have two different approaches. The first is theoretical in nature. We can construct a mathematical model of the system. By assumption, this model will be a linear ODE with constant coefficients from which F(s) is readily extracted:

$$F(s) = \frac{K}{s^n + C_{n-1}s^{n-1} + \ldots + C_1 s + C_o}$$

$$= \frac{\alpha\beta\gamma\ldots}{(s+\alpha)(s+\beta)(s+\gamma)\ldots} \qquad (9.6)$$

The second approach is experimental in spirit. Apply a unit impulse function at the inlet to the system: $a_{in}(t) = \delta(t)$, $A_{in}(s) = 1$. Then

$$F(s) = A_{out}(s) \qquad (9.7)$$

Equation 9.7 shows that the **transfer function is the output response to a unit impulse.**

## 9.1.1 Systems in Series

The block diagram below shows two systems in series

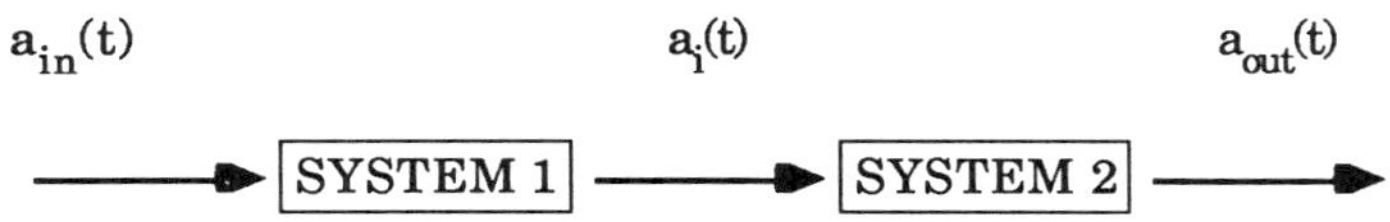


where $a_i$ is the intermediate, between-systems response. The above is a time domain representation. It is more common to show it in the Laplace domain:

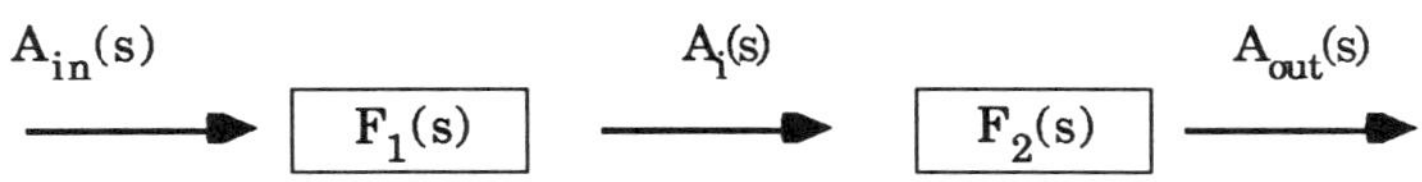


Applying Equation 9.4 to the first system gives $A_i(s) = F_1(s)A_{in}(s)$. Now $A_i(s)$ is the input to System #2. Thus, $A_{out}(s) = F_2(s)A_i(s)$. For the overall series system we have

$$\frac{A_{out}(s)}{A_{in}(s)} = F_1(s)F_2(s) = F_c(s) \qquad (9.8)$$

where $F_c$ is the transfer function for the combined series system. In block diagram form we can illustrate the series combination as

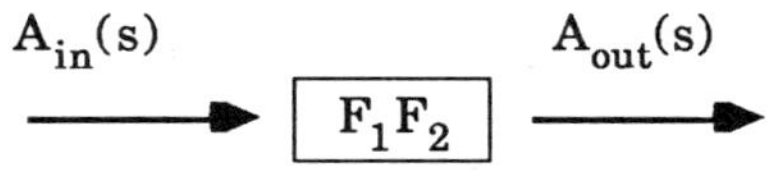


This result is easily generalized to N systems in series

$$F_c = F_1F_2...F_N = \prod_{j=1}^{N} F_j \tag{9.9}$$

## 9.1.2 Convolution

The multiplicative form of Equation 9.8 recalls our results for convolution in Section 4.39.

$$A_{out}(s) = F(s)\, A_{in}(s) = \pounds[f(t)*a_{in}(t)] \tag{9.10}$$

where

$$f(t) = \pounds^{-1}[F(s)] \tag{9.11}$$

is the **time domain transfer function** for the system. The time domain equivalent of Equation 9.10 is

$$a_{out}(t) = \int_0^t f(t-\tau)a_{in}(\tau)d\tau = \int_0^t f(\tau)a_{in}(t-\tau)d\tau \tag{9.12}$$

It will become apparent that any operation conducted in the Laplace domain can also be done in the time domain. Typically, the Laplace domain calculations are easier; but there are exceptions to this generalization. The time domain transfer function for a flow system is known as the **residence time density function** and has wide applications in chemical engineering. These will be explored in Chapter 10.

The time domain analogue of Equation 9.7 is

$$f(t) = a_{out}(t) \text{ when } a_{in}(t) = \delta(t) \tag{9.13}$$

The time domain analogue of Equation 9.8 is

$$f_c(t) = \int_0^t f_1(t-\tau)\, f_2(\tau)d\tau = \int_0^t f_1(\tau)\, f_2(t-\tau)d\tau \tag{9.14}$$

where $f_c(t)$ is the time domain transfer function for the two systems in series.

**EXAMPLE 9.1** Find $F_c$ and $f_c$ for two first order systems in series. Find the response of the composite system to an inlet step change, U(t).

**Solution:** The transfer functions are

$$F_1(s) = \frac{\alpha}{s + \alpha} \qquad F_2(s) = \frac{\beta}{s + \beta} \tag{i}$$

Using Equation 9.8,

$$F_c(s) = \frac{\alpha\beta}{(s + \alpha)(s + \beta)} \tag{ii}$$

Corresponding to Equation i are the time domain transfer functions

$$f_1(t) = \alpha e^{-\alpha t} \qquad f_2(t) = \beta e^{-\beta t} \tag{iii}$$

The time domain counterpart of Equation ii can be found by inversion. This gives

$$f_c(t) = \alpha\beta \left[\frac{e^{-\alpha t} - e^{-\beta t}}{\beta - \alpha}\right], \quad \alpha \neq \beta \tag{iv}$$

or

$$f_c(t) = \alpha^2 t e^{-\alpha t}, \qquad \alpha = \beta \tag{v}$$

An alternative to inverting Equation ii is to convolute $f_1$ and $f_2$:

$$f_c = f_1 * f_2 = \int_0^t \alpha e^{-\alpha(t-\tau)} \beta e^{-\beta\tau} d\tau$$

$$= \alpha\beta e^{-\alpha t} \int_0^t e^{(\alpha-\beta)\tau} d\tau$$

If $\alpha = \beta$, this integral gives Equation v. Otherwise it gives Equation iv.

For a unit step input, $A_{in}(s) = s^{-1}$. For this input,

$$A_{out}(s) = F_c A_{in} = \frac{\alpha\beta}{s(s+\alpha)(s+\beta)}$$

Convolution can be used as an alternative to inversion:

$$a_{out} = f_c * a_{in}$$

For the case of $\alpha = \beta$, $f_c$ is obtained from Equation v above. Thus,

$$a_{out} = \int_0^t \alpha^2(t-\tau)e^{-\alpha(t-\tau)}\,U(\tau)\,d\tau$$

$$= \alpha^2 te^{-\alpha t}\int_0^t e^{a\tau}d\tau - \alpha^2 e^{-\alpha\tau}\int_0^t \tau\, e^{d\tau}d\tau$$

$$= 1 - (\alpha t + 1)\,e^{-\alpha t}$$

---

## 9.2 NETWORK ANALYSIS FOR ELECTRICAL SYSTEMS

Section 9.1 considered composite systems consisting of two or more subsystems in series. We will now treat more complicated situations such as systems in parallel or systems with feedback. To do this, two new concepts need to be introduced, that of a **summation point** and that of a **splitting point**. For a summation point we draw the following diagram

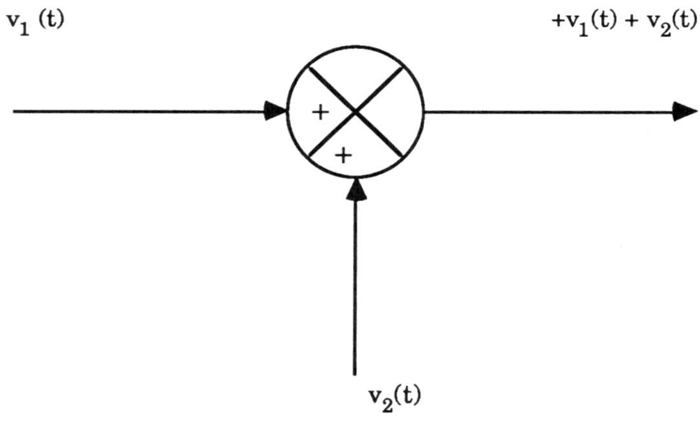


The device in the circle is a **voltage adder**. It gives the algebraic sum of the two input voltages, $v_1$ and $v_2$. The method of summation is indicated by the signs within the circle. For example, the following summation point reverses the sign on $v_1$.

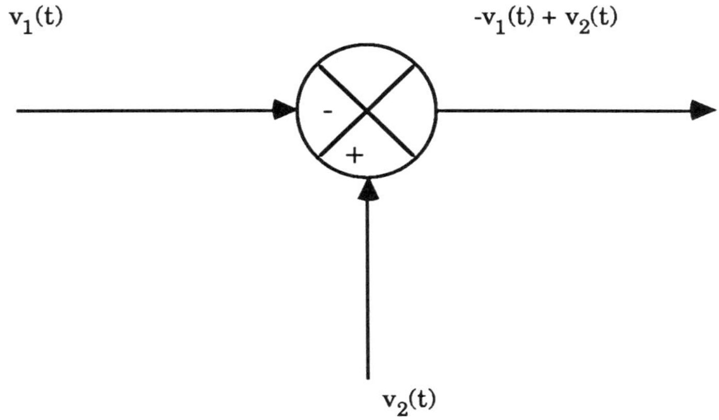


There are four different kinds of summation points. These different kinds depend on the signs within the circle: ++, +-, -+, --. There is only one kind of **splitting point**:

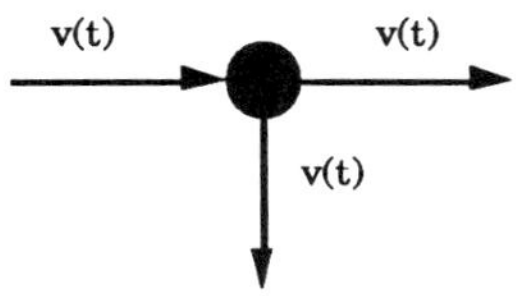


The voltage signal is split, without attenuation, and is passed along to any downstream devices.

The voltage adder is an analogue device that produces an output signal equal to the algebraic sum of the input voltages. Note that this form of addition is not the natural result of connecting two voltage sources at a common point. Voltage addition requires a special circuit to perform the addition.

**EXAMPLE 9.2** Find the transfer function for two subsystems in parallel.

**Solution:** The block diagram is

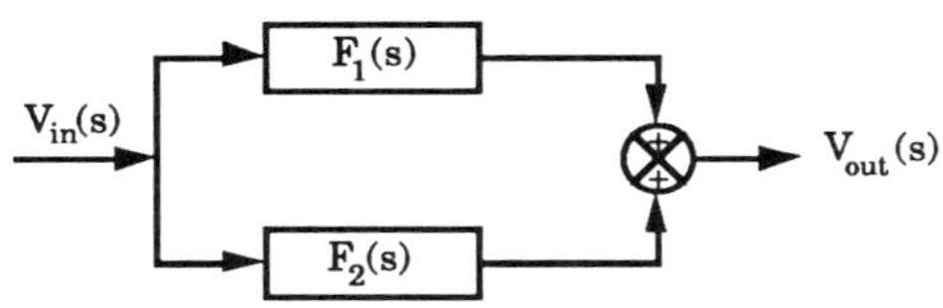


This situation has a splitting point followed by a summation point. Note that Laplace domain representation of splitting points and summation points is directly analogous to the time domain representation.

Using the definition of a splitting point, the block diagram becomes

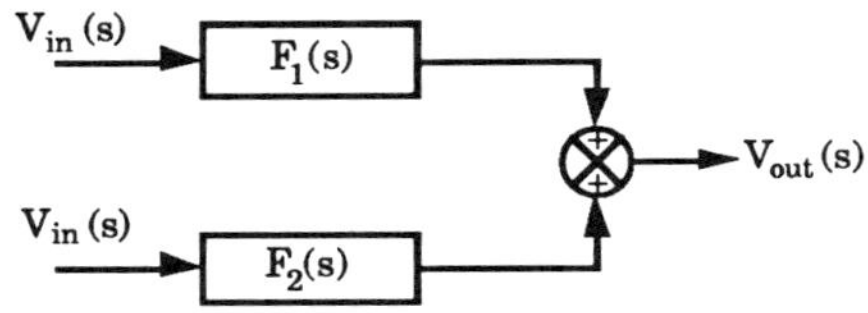


We use the property of transfer functions to obtain

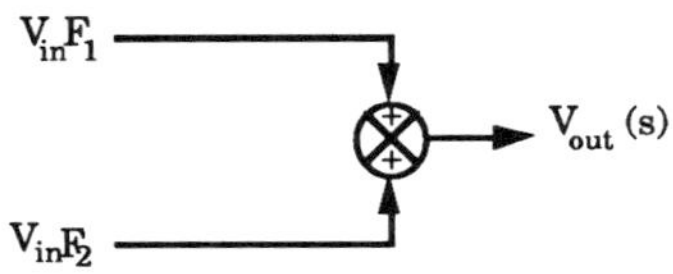


The definition of a summation point gives

$$V_{in}(s)F_1(s) + V_{in}(s)F_2(s) = V_{out}(s)$$

or, using Equation 9.5,

$$\frac{V_{out}(s)}{V_{in}(s)} = F_c(s) = F_1(s) + F_2(s)$$

---

The concepts of summation points and splitting points are not natural for concentrations in flow systems, and we defer consideration of such systems until Section 9.3. Thus, this section uses the symbols v(t) and V(s) rather than a(t) and A(s). Voltages can follow the rules of summation and splitting given above (although a special device is required for the addition), and these rules have been found useful in the analysis of control systems. We can also regard v(t) as a voltage **signal** which requires only one lead for its transmission (the other lead is ground) and which obeys the summation and splitting conventions defined above. Thus the various diagrams in this chapter are commonly called **signal flow diagrams**.

**EXAMPLE 9.3** Find the transfer function for a system with **instantaneous positive feedback** as indicated below.

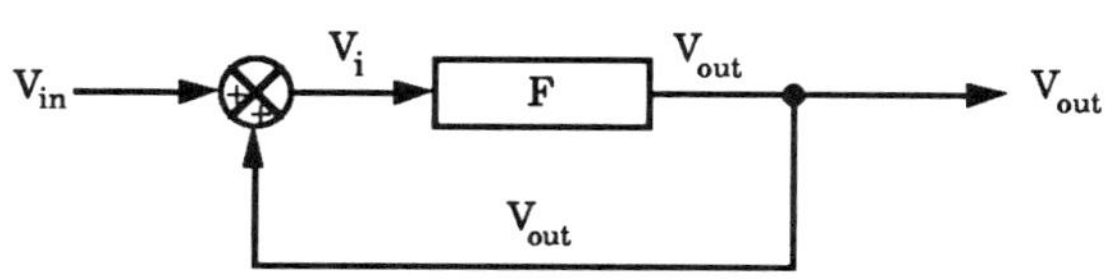

**Solution:** The intermediate voltage after the summation point is $V_i$ (in transform space). It is given by

$$V_i = V_{in} + V_{out}$$

Here, the **feedback signal** is just $V_{out}$ since we assume no lag of any kind between the splitting point and the summation point. The output from the system is $FV_i$ where

$$FV_i = F(V_{in} + V_{out}) = V_{out}$$

or

$$V_{out} = \frac{FV_{in}}{1 - F}$$

The transfer system for the feedback system is thus

$$F_c(s) = \frac{F}{1 - F}$$

---

We have three elementary concepts for block diagrams: systems in series, summation points, and splitting points. Systematic application of these basic concepts allows the analysis of composite systems of arbitrary complexity. The general approach is best illustrated by examples.

**EXAMPLE 9.4** Find the transfer function for the following system.

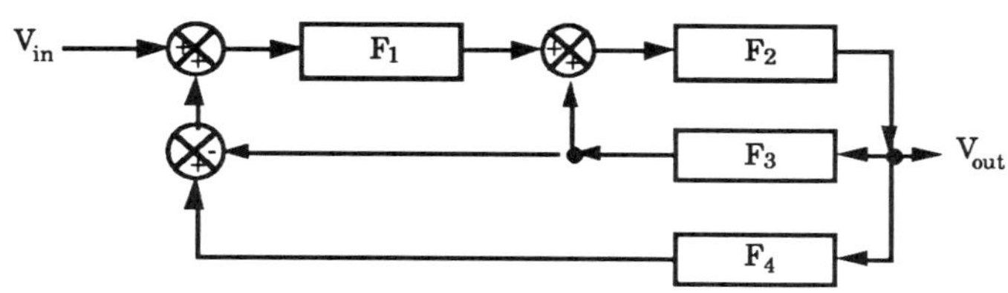


**Solution:** The overall system contains four subsystems arranged in a complex network of feedback loops. The first step in analyzing such a system is to label the various signals. One labeling scheme is shown below:

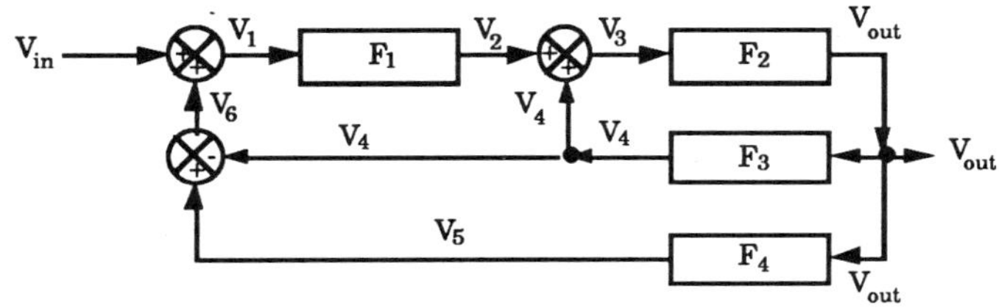


Note that the rules for splitting points have been applied directly. We now write equations for the various inline elements and summation points. For the inline elements we have

$$V_2 = F_1 V_1$$

$$V_{out} = F_2 V_3$$

$$V_4 = F_3 V_{out}$$

$$V_5 = F_4 V_{out}$$

For the summation points we have

$$V_{in} + V_6 = V_1$$

$$V_2 + V_4 = V_3$$

$$-V_4 + V_5 = V_6$$

There are seven equations and seven unknown transforms, $V_{out}$, $V_1$, ... $V_6$. We eliminate $V_1$ through $V_6$ to obtain the overall transfer function for the composite system:

$$F_c = \frac{V_{out}}{V_{in}} = \frac{F_1 F_2}{1 - F_2 F_3 + F_1 F_2 F_3 - F_1 F_2 F_4}$$

---

### 9.2.1 Feedback Systems

A general feedback system is shown below.

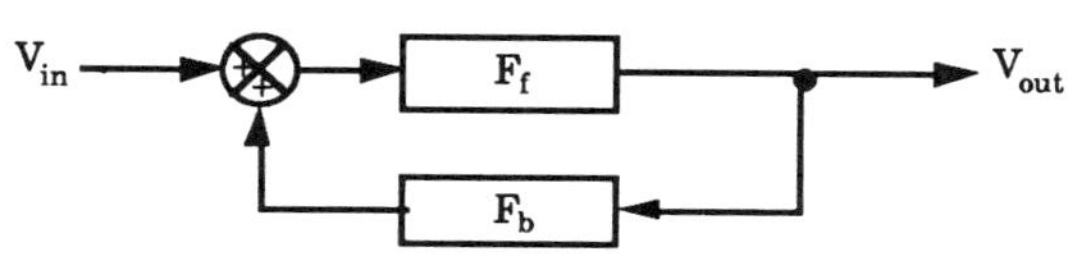

The transform for this composite system is

$$\frac{V_{out}}{V_{in}} = F_c = \frac{F_f}{1 - F_f F_b} \tag{9.15}$$

The sign in the denominator of Equation 9.15 depends on the way the summation point is operated. Since voltages and transfer functions can be signed quantities (that is, either positive or negative), it is possible to use only the ++ form of summation point without loss of generality. Negative feedback can be represented in any of the following ways:

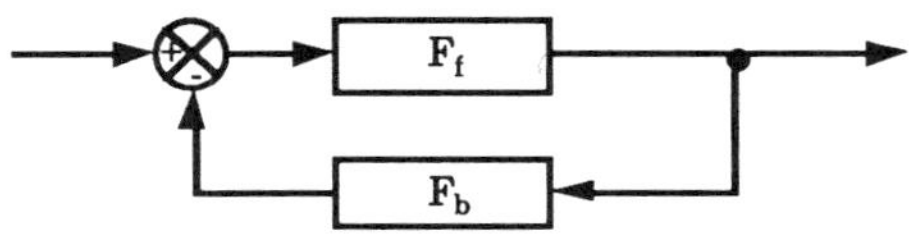


or

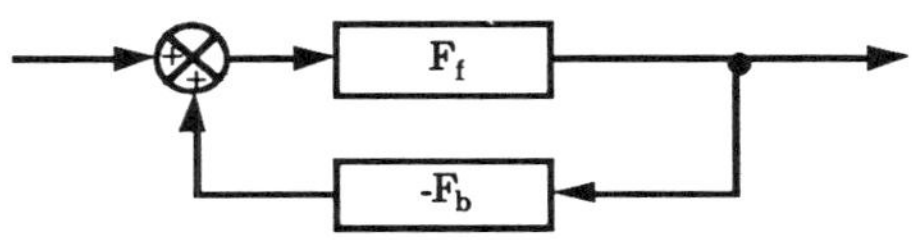


or

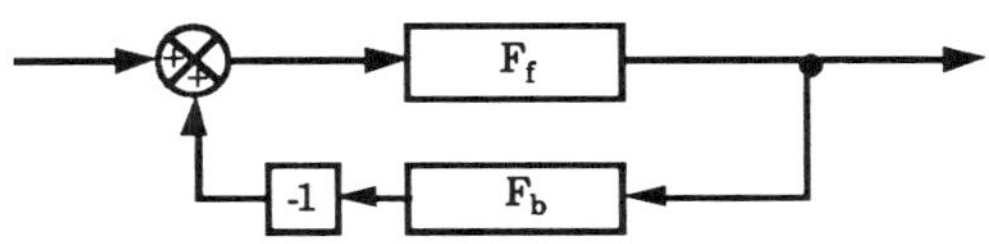


The last of these possibilities inserts a subsystem having transfer function F(s) = -1 into the feedback loop. [The time domain transfer function corresponding to F(s) = -1 is f(t) = -δ(t).] Regardless of how it is represented, the transfer function for a system with negative feedback is

$$F_c = \frac{F_f}{1 + F_f F_b} \tag{9.16}$$

**EXAMPLE 9.5** Determine the composite transfer function and the unit step response for the following feedback system.

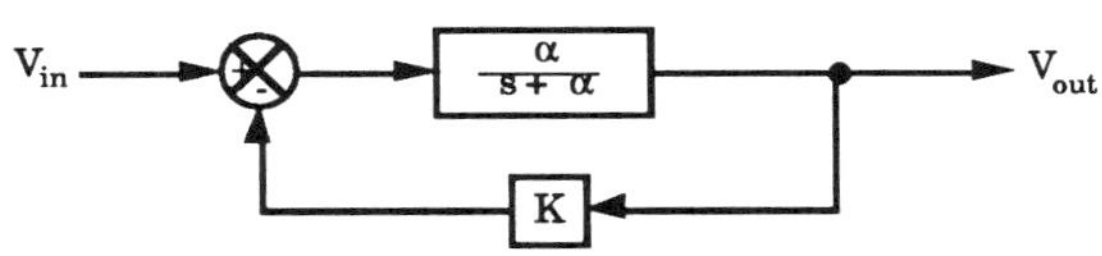

**Solution:** This is a straightforward application of Equation 9.16. $F_f$ corresponds to a first order system. $F_b = K$ corresponds to a simple **gain**. The composite transfer function is

$$F_c = \frac{\alpha}{s + \alpha + K\alpha}$$

The time domain transfer function (or equivalently, the impulse response function) is

$$f_c(t) = \alpha e^{-(1+K)\alpha t}$$

The response of the system to a unit step input is

$$V_{out} = F_c V_{in} = \frac{F_c}{s} = \frac{\alpha}{s(s+\alpha+K\alpha)}$$

or

$$v_{out}(t) = \frac{1-e^{-(1+K)\alpha t}}{1 + K} \qquad \text{(i)}$$

For K > 0, the feedback is negative, and the steady state response of the feedback system will be less than the **open loop** response for which K = 0. For 0 > K, feedback is positive. For 0 > K > -1 the pole of the composite system

$$s = -(1+K)\alpha$$

remains negative but is smaller in magnitude than the pole in the open loop system, $s = -\alpha$. In this region, the system continues to achieve a steady state when subjected to an input step function, but the magnitude of the response is larger than it would be without feedback. At K = -1, the pole vanishes. The step input response becomes

$$V_{out}(s) = \frac{\alpha}{s^2}, \; v_{out}(t) = \alpha t$$

for the special case where K = -1. For -1 > K, the pole is positive. The output no longer achieves a steady state but grows exponentially in time.

---

Example 9.5 showed the effect of feedback on a first order system. Excessive positive feedback causes the system to behave much like a ramjet engine or a turbo-charged car. The faster it goes the faster it goes. Real systems will saturate or break. Mathematical systems will diverge to infinity.

Negative feedback mitigates the response of a system compared to its **open-loop** (no feedback) response. Negative feedback is used in **feedback control systems**. Such systems are discussed briefly in Section 9.2.2 and extensively in courses on process control. In the context of control systems design, positive feedback is generally undesirable. Negative feedback is generally desirable but, like most good things, it too can be carried to excess.

**EXAMPLE 9.6** Determine the composite transfer function and the unit step response for a second order system with feedback. Assume the time constants are real and unequal.

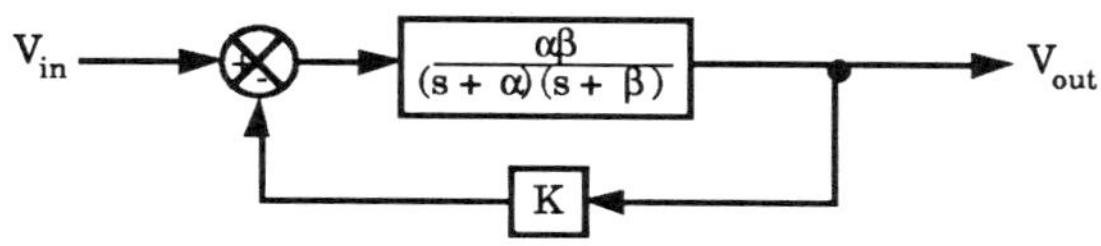


**Solution:** Equation 9.16 gives

$$F_c = \frac{\alpha\beta}{s^2 + (\alpha+\beta)s + (1+K)\alpha\beta}$$

The two poles of the composite transfer function are

$$s_1, s_2 = \frac{(-\alpha-\beta) \pm \sqrt{\alpha^2 + [2-4(1+K)]\alpha\beta + \beta^2}}{2}$$

The sum of the two roots is constant and independent of K:

$$s_1 + s_2 = -\alpha - \beta$$

The allocation of this constant sum between the roots depends on K. For $K = 0$, we get the open loop response with $s_1 = -\alpha_1$, $s_2 = -\beta$. For $K < 0$, one pole grows and the other decreases, but both poles remain negative in the range $-1 < K < 0$. At $K = -1$, one pole becomes zero, and the system becomes unstable. This is another example of destabilization due to positive feedback. [Note that the feedback is positive for $K < 0$ because we are using a negative summation for the feedback loop. According to conventional usage, a **negative gain** ($K < 0$) gives positive feedback.] For $K < -1$, one pole is positive; one is negative; and the system is unstable.

Now consider the case of $K > 0$ so that the feedback is negative. Both roots are negative and, at least for moderately positive K, the system achieves a steady state. However, at

$$K_{crit} = \frac{(\alpha + \beta)^2}{4\alpha\beta} - 1$$

the roots become equal:

$$s_1 = s_2 = \frac{-(\alpha + \beta)}{2}$$

This marks the transition from real to complex roots. For $K > K_{crit}$, the system will overshoot its steady state value and will oscillate. The oscillation will always be damped since the real portion of the roots will remain constant at $-(\alpha + \beta)/2$. The frequency of the oscillation will increase with increasing K.

The time domain response to an inlet step change of unit height can be found from

$$V_{out}(s) = \frac{\alpha\beta}{s(s-s_1)(s-s_2)}$$

This inverts to

$$v_{out}(t) = U(t) + \frac{s_2 e^{s_1 t}}{s_1 - s_2} + \frac{s_2 e^{s_1 t}}{s_2 - s_1} \qquad \text{(i)}$$

Equation i can be used to obtain the time domain response in numerical or graphical form. However, it is easier to use the simulation techniques of Chapter 12. Our presentation of these results is given in Example 12.3.

---

## 9.2.2 Feedback Control

Negative feedback provides the basis for automatic control. Suppose it is desired to hold a dependent variable, say the outlet temperature from a water heater, at some fixed value which we will call the **setpoint**. If the actual temperature is too low, the heating system is turned up. If it is too high, the heating system is turned down. The error signal

$$e(t) = T_{out} - T_{set} \qquad (9.17)$$

generates a control action in the opposite direction. Thus, if $e(t) > 0$, heat will be removed, and if $e(t) < 0$, heat will be added. This is negative feedback.

The specific control action depends on the choice of a **control variable** and on the nature of the **controller**. Suppose the water is heated by a steam coil held at temperature $T_C$. Then one possible control variable is $T_C$ and one possible control algorithm is

$$T_c = T_c(0) - K_P e(t) \qquad (9.18)$$

where $K_P$ is a constant called the **proportional gain**. Equation 9.18 corresponds to a **proportional-only** controller since the control action is strictly proportional to the error signal. A more complicated controller is

$$T_c = T_c(0) - K_P e(t) - K_I \int_0^t e(t)dt - K_D \frac{de}{dt} \tag{9.19}$$

where we have added **integral** (or **reset**) and **derivative** actions. In this scheme, $K_P$ is dimensionless, $K_I$ has units of reciprocal time, and $K_D$ has units of time. Note the minus sign before the e(t) terms. All of these actions represent negative feedback of the error signal to the control variable.

A few basic concepts of feedback control will be illustrated through examples.

**EXAMPLE 9.7** Devise a proportional controller for the stirred tank water heater illustrated below. Assume the flow rate through the heater, Q, is constant. Find the response of $T_{out}$ to a change in inlet water temperature.

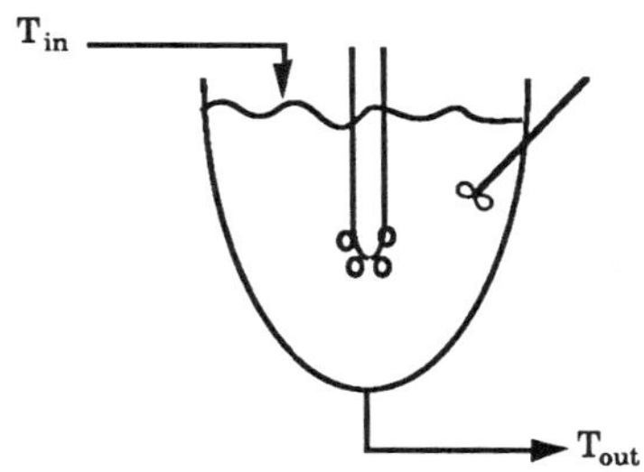


**Solution:** The first step is to model the system. The methods of Chapter 7 give

Sensible Heat In + Heat from Coil = Sensible Heat Out + Accumulation of Heat

$$\rho c Q T_{in} + hA_c(T_c - T_{out}) = \rho c Q T_{out} + \rho c V \frac{dT_{out}}{dt}$$

or

$$\frac{dT_{out}}{d\theta} + (1 + H)\, T_{out} = T_{in} + HT_c \tag{i}$$

where $\theta = tQ/V$ and $H = hA_c/\rho cQ$. Equation i together with an initial condition on $T_{out}$ constitutes the system model. The system inputs are $T_{in}$ and $T_c$. If $T_{in}$ changes, we want to compensate for the change by manipulating $T_c$.

The error signal is defined as

$$e = T_{out} - T_{set} \tag{ii}$$

and the control action is given by

$$T_c = T_c(0) - K_P e \tag{iii}$$

Equation iii causes negative feedback control. If $T_{out}$ is too high, then $e > 0$ and Equation iii reduces the heat input by reducing $T_c$. If $T_{out}$ is too low, Equation iii increases the heat input by increasing $T_c$. A real water heater would need physical devices such as a thermocouple to measure $T_{out}$, a controller to perform Equations ii and iii, and a pressure regulator to adjust $T_c$. These devices will have dynamics of their own which can be important in some control problems. In the current problem, it is reasonable to suppose that the measurement and control hardware have time constants much smaller than that for the stirred tank water heater. Thus, we suppose they achieve their functions instantaneously so that the system is completely modeled by Equations i, ii, and iii. These equations contain three time-dependent unknowns, $T_{out}$, e, and $T_c$. They are solved simultaneously. Laplace transformation gives

$$s\tau_{out} - T_{out}(0) + (1+H)\tau_{out} = \tau_{in} + H\tau_c$$

$$E = \tau_{out} - \tau_{set}$$

$$\tau_c = \frac{T_c(0)}{s} - K_P E$$

where $\tau = £[T]$. Solution for E gives

$$E = \frac{\tau_{in} + T_{out}(0) + \dfrac{HT_c(0)}{s} - \tau_{set}[s+1+H]}{s + 1 + H + K_P H} \tag{iv}$$

This result is quite general in the sense that it allows for set point changes with $T_{set} = T_{set}(t)$, demand changes with $T_{in} = T_{in}(t)$, and initial offsets with $T_{out}(0) \neq T_{set}$. We will illustrate the response of the system to a step change in the inlet water temperature. Prior to the step change, suppose the system was at steady state with $e = 0$ and $dT_{out}/dt = 0$. Then Equation i gives

$$(1+H)\,T_{out}(0) = T_{in}(0) + HT_c(0)$$

This result can be used to eliminate $T_c(0)$ from Equation iv.

We suppose $T_{set}$ to be constant so that

$$\tau_{set} = \frac{T_{set}}{s} = \frac{T_{out}(0)}{s}$$

For the inlet step change we have

$$T_{in}(t) = T_{in}(0) + \Delta T_{in} U(t)$$

or

$$\tau_{in} = \frac{T_{in}(0) + \Delta T_{in}}{s}$$

Substituting into Equations iv gives

$$E = \frac{\Delta T_{in}}{s(s + 1 + H + K_P H)} \qquad \text{(v)}$$

The inverse is

$$e(\theta) = \frac{\Delta T_{in}[1 - e^{-(1+H+K_P H)\theta}]}{1 + H + K_P H} \qquad \text{(vi)}$$

The original steady state had e = 0 by assumption. Equation vi approaches a new steady state, but there is now a permanent error called the **offset**:

$$e(\infty) = \frac{\Delta T_{in}}{1 + H + K_P H} \qquad \text{(vii)}$$

The open loop response ($K_P = 0$) gives a larger error than that of Equation vii, so the proportional controller has done some good. In theory, the offset can be made arbitrarily small by using large values for $K_P$. In practice, a real controller will have dynamics of its own and will cause oscillations if $K_P$ is too large. A better approach is to add an integral term to the controller.

---

**EXAMPLE 9.8** Modify the coil temperature controller to include integral (reset) action.

**Solution:** The only change in Example 9.7 is the control action. Now

$$T_c = T_c(0) - K_P e - K_I \int_0^t e\,dt$$

and

$$\tau_c = \frac{T_c(0)}{s} - K_P E - \frac{K_I E}{s}$$

The solution for E becomes

$$E = \frac{\tau_{in} + T_{out}(0) + \dfrac{HT_c(0)}{s} - (s + 1 + H)\,\tau_{set}}{s + 1 + H + K_P H + K_I H/s}$$

$$= \frac{[\tau_{in} + T_{out}(0) - (s + 1 + H)\tau_{set}]s + HT_c(0)}{s^2 + (1 + H + K_P)s + K_I H}$$

Suppose the system was at an initial steady state but that a step change in $T_{in}$ occurred at $t = 0$. As in Example 9.7, we use the conditions at the initial steady state to obtain

$$\tau_{out} = \frac{T_{out}(0)}{s} + \frac{\Delta T_{in}}{[s^2 + (1 + H + HK_P)s + HK_I]}$$

Applying the final value theorem gives

$$T_{out}(\infty) = \lim_{s \to 0} s\tau_{out} = T_{out}(0)$$

Thus, the system returns to its initial steady state. The integral action ($K_I > 0$) eliminates the permanent offset found with proportional-only control.

The poles of $\tau_{out}$ are

$$s_1, s_2 = \frac{-(1+H+HK_P) \pm \sqrt{(1+H+HK_P)^2 - 4HK_I}}{2}$$

The poles will be complex if $4HK_I > (1 + H + HK_P)^2$. Thus, integral action can cause oscillations (**overshoot**). Some overshoot is generally desirable for fast response to process upsets, but that is a topic for a course in process control where optimal choices for $K_p$, $K_I$, and $K_D$ are considered.

---

## 9.3 NETWORK ANALYSIS FOR FLOW SYSTEMS

Network analysis can be applied to chemical concentrations in flow systems, but the rules for summation and splitting points are different. At a **summation point**, the flow rates are summed, but the concentrations are flow averaged.

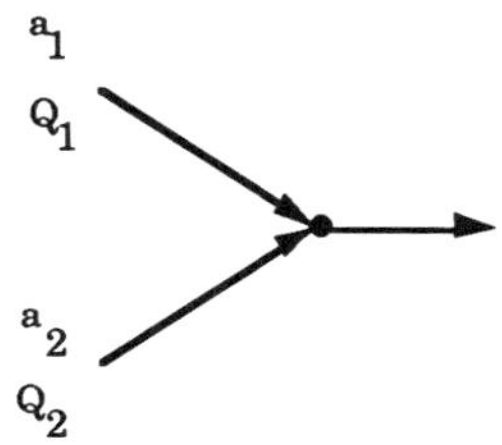


$$a_{out} = \frac{a_1 Q_1 + a_2 Q_2}{Q_1 + Q_2}$$

$$Q_{out} = Q_1 + Q_2$$

At **splitting points**, the flow is divided, but the concentrations are unchanged:

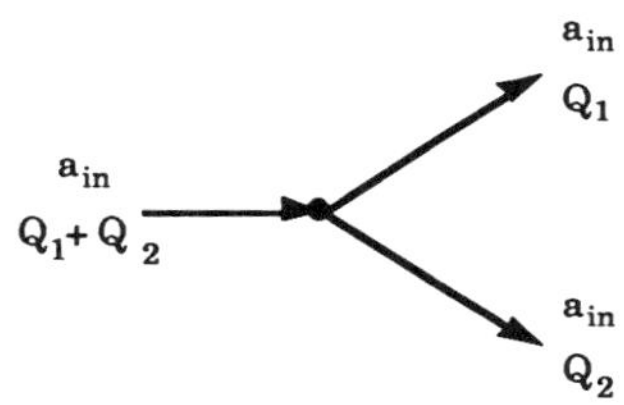


Feedback loops can be used in flow systems, but the feedback is always positive. Thus, we use a simple dot to represent both summation and splitting. The corresponding physical devices are pipe tees and crosses.

The block diagram for flow systems in parallel is

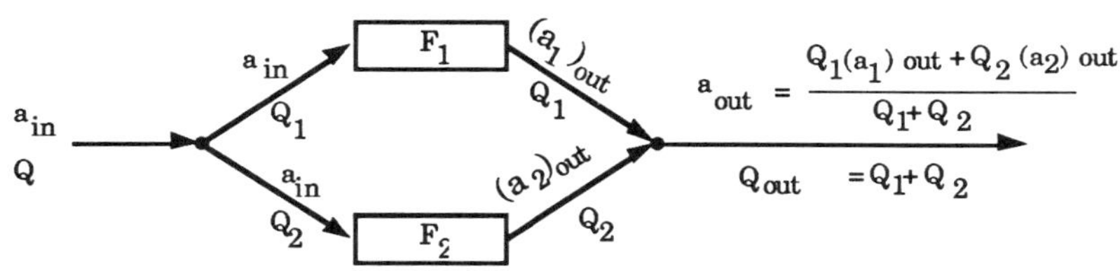


---

**EXAMPLE 9.9** Apply network analysis to the following flow system.

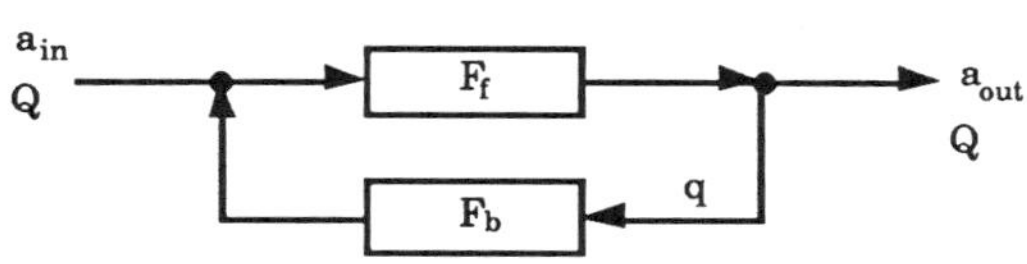

**Solution:** Redraw the block diagram to label the interior flows and concentrations

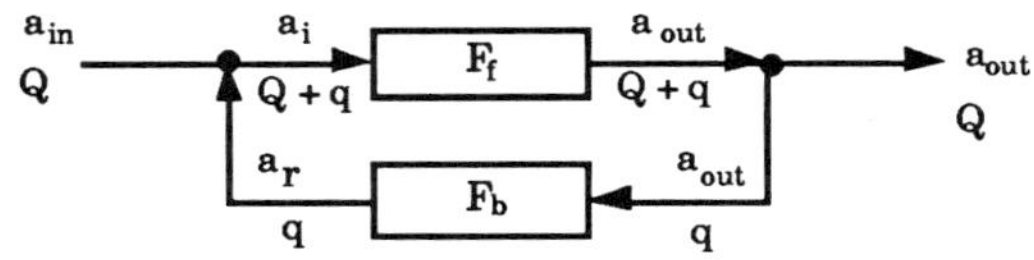


Here, the various flow rates are known, Q, q, and Q+q; but there are three unknown concentrations, $a_i$, $a_{out}$, and $a_r$. We suppose $a_{in}(t)$ is known and that the system is initially relaxed. Then $A_{out} = A_i F_f$ and $A_r = A_{out} F_b$ using the usual rules for transfer functions. Concentration averaging is the same in the Laplace domain as in the time domain.

Thus,

$$A_{out} = A_i F_f = \left[\frac{QA_{in} + qA_r}{Q + q}\right] F_f = \left[\frac{QA_{in} + qA_{out}F_b}{Q + q}\right] F_f \tag{9.20}$$

Solving for $A_{out}$ gives

$$\frac{A_{out}}{A_{in}} = F_c = \frac{F_f}{1 + [q/Q - (q/Q)F_b F_f]} \tag{9.21}$$

The term in brackets is due to the feedback loop. It is reminiscent of, but different than, positive feedback in an electrical circuit. (Compare Equation 9.15.)

---

**EXAMPLE 9.10** Determine the overall transfer function of two stirred tank reactors in series. Assume that reactant A is consumed by a first order reaction. The flow diagram is

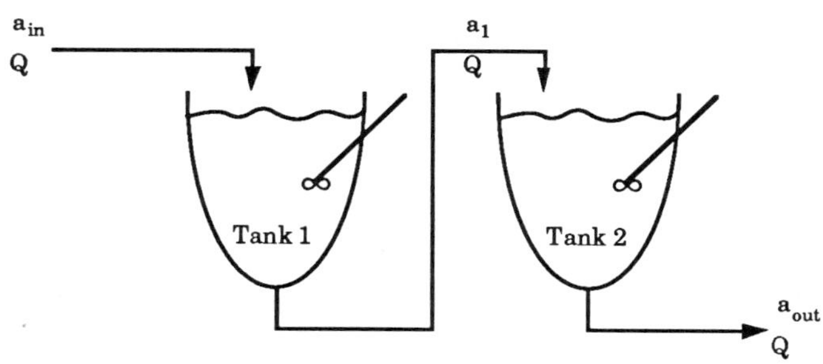

**Solution:** The methods of Chapter 2 give the following unsteady state material balance:

Inflow of A + Rate of Formation of A = Outflow of A + Accumulation of A.

For Tank #1,

$$Qa_{in} - V_1k_1a_1 = Qa_1 + V_1\frac{da_1}{dt}$$

or

$$t_1\frac{da_1}{dt} + (1 + k_1t_1)a_1 = a_{in}$$

where $t_1 = V_1/Q$. For Tank #2,

$$t_2\frac{da_2}{dt} + (1 + k_2t_2)a_2 = a_1$$

where $t_2 = V_2/Q$.

In block diagram form,

The overall transfer function is

$$\frac{A_{out}}{A_{in}} = F_c = \frac{1}{[(s+k_1)t_1 + 1]\,[(s+k_2)t_2 + 1]}$$

The above result governs the transient operation of two stirred tank reactors in series. A first order reaction was assumed (Laplace techniques do not work for other reaction orders), but the volumes and rate constants can be different in the two reactors. The assumption of an initially relaxed system, used throughout this chapter, supposes the tanks to have been initially free of reactant.

To find the steady state response of the system, set $a_{in} = a_oU(t)$ and let $t \to \infty$. The final value theorem gives the desired result without need for inversion.

$$a_{out}(\infty) = \lim_{s\to 0} sA_{out} = sA_{in}F_c$$

$$= \lim_{s\to 0} s[\frac{a_0}{s}]\,F_c = \frac{a_0}{(1 + k_1t_2)\,(1 + k_2t_2)}$$

**EXAMPLE 9.11** Apply feedback (**recycle**) to the stirred tanks of Example 9.10. The system to be studied is

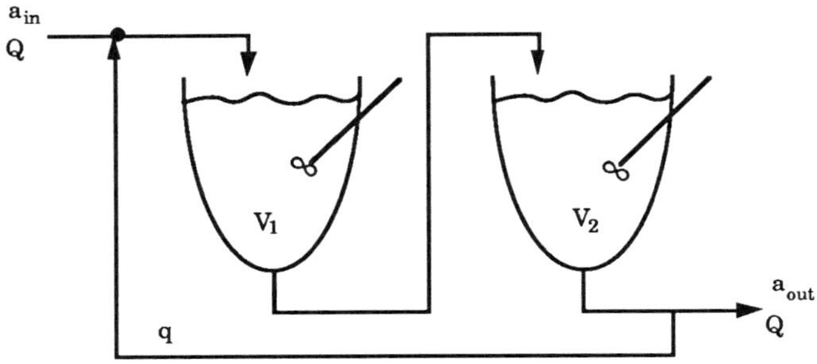


**Solution:** It is important to recognize that recycle, by increasing the flow through the tanks, changes the values for $t_1$ and $t_2$. Thus,

$$t_1 = \frac{V_1}{Q + q}, \qquad t_2 = \frac{V_2}{Q + q}$$

where $q > 0$. (The case with $q < 0$ is physically possible. It corresponds to **bypassing** rather than **recycle**.) In block diagram form,

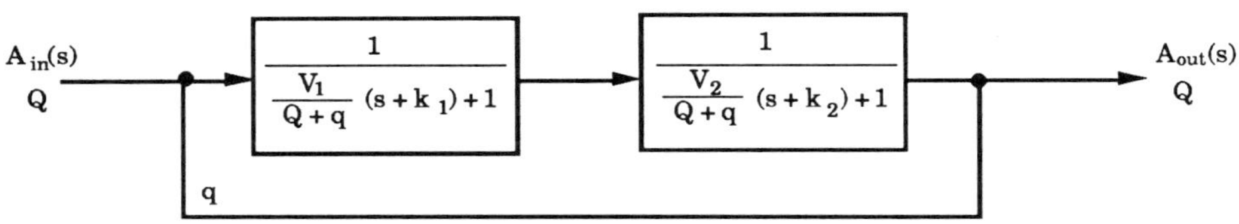


In this example, the recycle is instantaneous with $F_b = 1$. The overall transfer function follows from Equation 9.21 with $F_b = 1$ and $F_f = F_1 F_2$.

$$\frac{A_{out}}{A_{in}} = F_c = \frac{F_1 F_2}{1 + \frac{q}{Q} - \frac{q}{Q} F_1 F_2}$$

where $F_1$ and $F_2$ have the functional forms shown in the block diagram. Algebra gives

$$\frac{A_{out}}{A_{in}} = F_c = \frac{Q(Q+q)}{(V_1 s + Q + q + k_1 V_1)(V_2 s + Q + q + k_2 V_2) - q(Q+q)} \tag{i}$$

The steady state output for this reactor configuration is

$$a_{out}(\infty) = \frac{a_o Q\,(Q+q)}{(Q + q + k_1V_1)\,(Q + q + k_2V_2) - q\,(Q+q)}$$

---

**EXAMPLE 9.12** Consider two equal volume stirred tanks in series, without reaction but with recycle, as shown below.

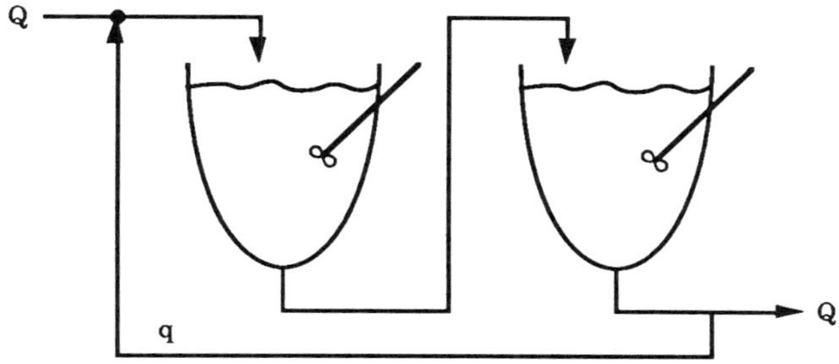


Will recycle cause instabilities (positive poles) or oscillations (complex poles)?

**Solution:** We start with Equation i of the previous example and specialize it with $V_1 = V_2 = V/2$, $k_1 = k_2 = 0$. The result is

$$\frac{A_{out}}{A_{in}} = F_c = \frac{Q(Q+q)}{(\frac{V}{2}\,s + Q + q)^2 - q(Q+q)}$$

We test the roots of

$$(\frac{V}{2}\,s + Q + q)^2 - q(Q+q) = (\frac{V}{2}\,s + Q + q)^2 \ - q+(Q + q) = 0$$

These roots are

$$s_1, s_2 = \frac{-2[Q + q \pm \sqrt{q(Q+q)}]}{V}$$

Recycle implies $q > 0$. In this case the roots are real, unequal, and negative. Thus, recycle induces neither oscillations nor instabilities for this system of two tanks in series.

---

## PROBLEMS

9.1 Complete Example 9.1 by finding the response to a unit step change of two unequal, first order systems in series. Use convolution, rather than inversion to find this result.

9.2 Convolution integrals come in pairs. See Equation 9.12. You used one member of the pair to work Problem 9.1. Now repeat Problem 9.1 using the other member.

9.3 Find $F_C$ for the following voltage system.

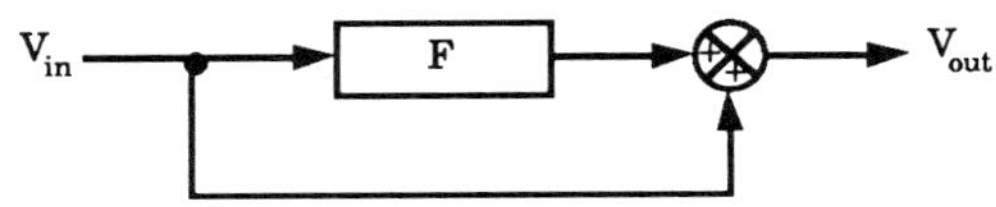


9.4 Find $F_C$ for the following flow system.

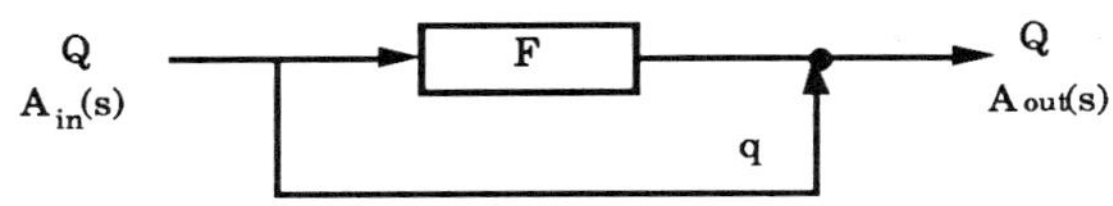


given that $Q > q > 0$.

9.5 Refer to Example 9.6.

(a) Verify that Equation i is indeed the time domain response to a unit step change.

(b) Determine the time domain response for the case of equal roots, $\alpha = \beta$.

Hint: Take the limit as $\alpha \rightarrow \beta$.

9.6 Suppose the water heater of Examples 9.7 and 9.8 is subject to a setpoint change where $T_{set}$ is changed by an amount $\Delta T_{set}$ at $t = 0$. Find the outlet response assuming a feedback controller with proportional and integral actions.

9.7 Suppose derivative action is added to the controller in Example 9.8. Find E(s).

9.8 Repeat Example 9.11 but reverse the flow direction in the bypass line, thus giving bypassing.

9.9 Consider the following flow system.

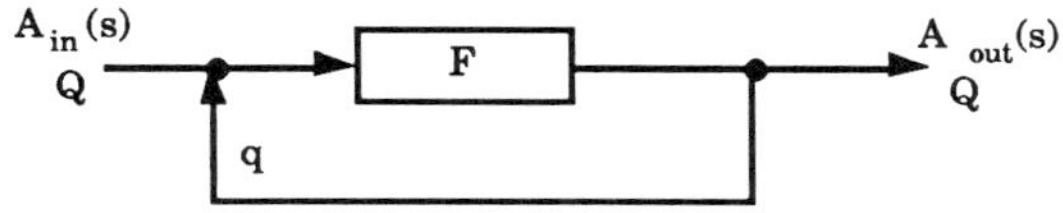


Suppose F corresponds to a single stirred tank with consumption of A by a first order reaction. How does recycle affect the steady-state performance of the reactor?

9.10 Refer to Equation 9.2. Prove that a(t) is a conserved quantity.

$$\int_0^\infty a_{in}(t)dt = \int_0^\infty a_{out}(t)dt$$

# CHAPTER 10

# RESIDENCE TIME DISTRIBUTIONS

Flow systems have **particles** flowing through them. Typically, these particles will be molecules, but they can be larger entities. Whatever their size, each particle in a flow system will have a **residence time**. This is the time that the particle spends inside the boundaries of the system. The easiest way to envision **residence time** is to suppose each particle carries a time card. The card is punched when a particle enters the system. It is punched again when the particle leaves. The elapsed time between entry and exit is the residence time, t. (This book ignores the possibility of multiple entries and exits.)

Not all particles entering a system will experience the same residence time. There will be an average time, $\bar{t}$, which we have already encountered and which is called the **mean residence time**. It is an important property of steady state flow systems that

$$\bar{t} = I_p/Q_p \tag{10.1}$$

where $I_p$ is the number of particles inside the system boundaries and $Q_p$ is the flow rate of particles into and out of the system. For the case of constant particle density per unit volume of fluid,

$$\bar{t} = \frac{V}{Q} \tag{10.2}$$

where V is the system volume and Q is the volumetric flow rate.

In real flow systems, some particles will have a residence time less than average and others will have a residence time greater than average. To quantify this **distribution of residence times**, we define a **residence time density function**, f(t), as

$$f(t)dt = \text{fraction of particles leaving the system which experienced a residence time between t and t+dt} \tag{10.3}$$

The function f(t) can be quite general. It is defined only for positive values of t since residence time must be positive. Also, $f(t) \geq 0$ for all $t > 0$ since f(t)dt is a fraction. Finally, all particles will eventually leave the system. This leads to

$$\int_0^\infty f(t)dt = 1 \tag{10.4}$$

Equation 10.4 states that the cumulative fraction of particles leaving the system must sum to 1 if we consider all possible exit times.

Closely related to f(t) is the **residence time washout function**, w(t), defined as

$$w(t) = \text{fraction of particles leaving the system which experienced a residence time greater than t} \tag{10.5}$$

The functions f(t) and w(t) are related by

$$w(t) = \int_t^\infty f(t)dt = 1 - \int_0^t f(t)dt \tag{10.6}$$

The properties of w(t) include $w(0) = 1$ since all particles have a residence time greater than 0, and $w(\infty) = 0$ since all particles eventually leave the system. It is also true that w(t) is **monotonic decreasing** ($dw/dt \leq 0$). To see this, differentiate Equation 10.6 and note that $f(t) \geq 0$.

Typical f(t) and w(t) are shown below:

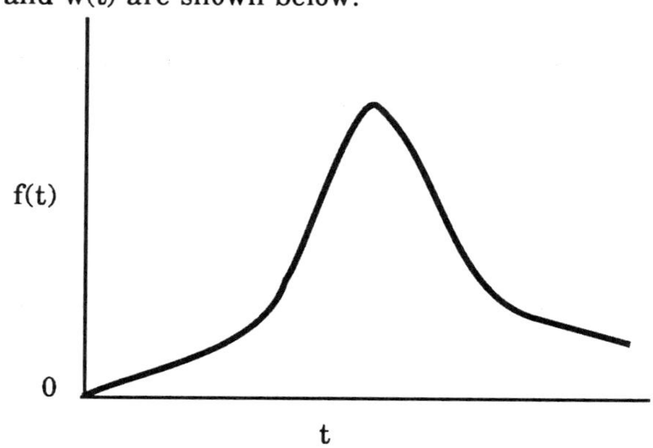

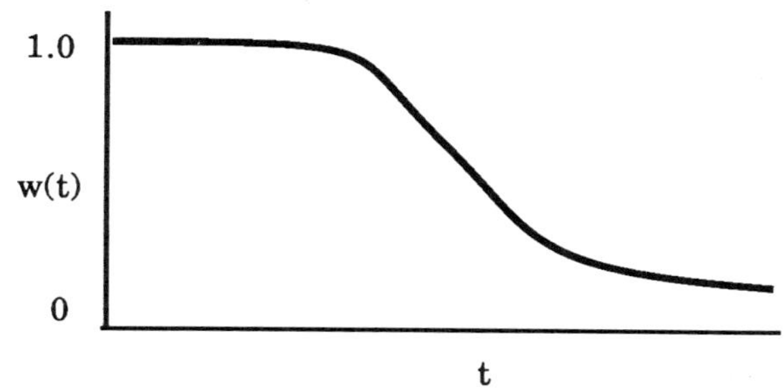


## 10.1 Stimulus-Response Techniques

The simplest way of measuring a residence time distribution is to inject a sharp burst of tracer particles into the system at time t = 0. Mathematically, we consider the input to be a unit impulse function, δ(t). The outlet concentration of tracer particles is monitored as a function of time. The experiment gives f(t) directly:

$$c_{out}(t) = f(t) \text{ when } c_{in}(t) = \delta(t) \qquad (10.7)$$

Equation 10.7 supposes the usual property of a delta function

$$\int_0^\infty \delta(t)\, dt = 1 \qquad (10.8)$$

When this is true, Equation 10.4 will also be true since the tracer particles are a conserved species (they do not react). In practice, experimentally measured concentrations are only proportional to absolute particle concentrations. This means that the outlet response must be scaled to ensure that Equation 10.4 is satisfied. The scaling is

$$f(t) = \frac{c_{out}(t)}{\int_0^\infty c_{out}(t)dt} \qquad (10.9)$$

In cases where a system model exists, f(t) can be found theoretically. It is the response of a flow system to a unit impulse. Thus,

$$f(t) = \mathcal{L}^{-1}[F(s)] \qquad (10.10)$$

where F(s) is the transfer function of the system.

Another experimental method for measuring residence time distributions is to inject a unit step change at the inlet. Suppose the step change is of the negative, turn-off variety. Then the washout function is obtained directly:

$$c_{out}(t) = w(t) \text{ when } c_{in}(t) = 1 - U(t) \tag{10.11}$$

In practice, concentration measurements must be scaled by $c_{out}(0)$ to ensure that $w(0) = 1$.

Other kinds of tracer input signals can be used to measure f(t) or w(t), but only impulse functions or step changes give the computationally simple results shown above.

Equation 10.6 gives the time domain relationship between f(t) and w(t). The Laplace domain counterpart of Equation 10.6 is

$$\mathcal{L}[w(t)] = W(s) = \frac{1-F(s)}{s} \tag{10.12}$$

**EXAMPLE 10.1** Suppose the transfer function of a system is $F(s) = \frac{1}{s+\alpha}$. Can this system represent the flow of nonreactive tracer molecules?

**Solution:** We invert F to obtain

$$f(t) = e^{-\alpha t}$$

as a possible residence time density function. This function is always positive so it can be a fraction, and we can consider it to be defined for positive times only. However, it fails to satisfy Equation 10.4 since

$$\int_0^\infty e^{-\alpha t}\, dt = \frac{1}{\alpha}$$

Thus, $F(s) = \dfrac{1}{s+\alpha}$ cannot represent the flow of inert tracer particles. Instead, the transfer function of a first order system must be

$$F(s) = \frac{\alpha}{s+\alpha}$$

The corresponding ODE for this first order system is

$$\frac{dc_{out}}{dt} + \alpha\, c_{out} = \alpha c_{in}$$

---

## 10.2 MEANS AND MOMENTS

A residence time distribution can be a deterministic function. This is the case for diffusion-free laminar flow. It can also be a probablistic function. This is the case for turbulence. Whether deterministic or probablistic, the language and mathematical tools of statistics and probability theory are useful for describing residence time distributions. Two important descriptors are the **mean** and **variance** of the distribution.

The mean residence time is defined as

$$\bar{t} = \int_0^\infty t f(t) dt \tag{10.13}$$

For the flow of inert tracer particles, it happens that $\bar{t} = I_p/Q_p$ or $\bar{t} = V/Q$ as given by Equations 10.1 or 10.2. Thus, there are two different ways of determining $\bar{t}$ experimentally:

1. Inject an impulse of tracer. Measure f(t) as the response function. Evaluate Equation 10.13.
2. Measure V and Q (or $I_p$ and $Q_p$). Calculate t using Equation 10.2 (or 10.1).

The two methods must agree. If they do not, at least one of the measurements is wrong. If V is unknown, it can be determined by an inert tracer experiment. This trick can be used to find the filled volume in a variable volume system: perform the tracer experiment to get $\bar{t}$, measure Q; and then calculate $V = Q\bar{t}$.

The mean residence time can also be calculated from the washout function:

$$\bar{t} = \int_0^\infty w(t) dt \tag{10.14}$$

When $\bar{t}$ is to be determined experimentally, step changes are preferred to impulses. Scaling the data is easier, and numerical integration of Equation 10.14 is more accurate than numerical integration of Equation 10.13.

The **variance** of the distribution is given by

$$\sigma_t^2 = \int_0^\infty (t-\bar{t})^2 f(t) dt = \int_0^\infty t^2 f(t) dt - (\bar{t})^2 \tag{10.15}$$

It is also given by

$$\sigma_t^2 = 2\int_0^\infty t\, w(t) dt - (\bar{t})^2 \tag{10.16}$$

Thus $\sigma_t^2$ can be calculated directly using either impulse-response data or washout data. Washout data are preferred for accuracy.

The variance of a distribution measures its broadness. If $\sigma_t^2 = 0$, all particles have the same residence time, t = t. If $\sigma_t^2 > 0$, there will be a spread of residence times. There is no upper limit on $\sigma_t^2$, and systems with $\sigma_t^2 = \infty$ have theoretical interest.

As defined above, $\sigma_t^2$ has units of [time$^2$]. A more useful descriptor of the shape of a distribution is the **dimensionless variance,**

$$\sigma^2 = \frac{\sigma_t^2}{(\bar{t})^2} \tag{10.17}$$

The numerical value for $\sigma^2$ provides a measure of **mixing** between particles which entered the system at different times. If $\sigma^2 \approx 0$, such mixing is small. If $\sigma^2 \approx 1$, there is considerable mixing. (The case of $\sigma^2 > 1$ is special, unusual in well-designed vessels, and will not be treated here. See, however, Example 10.5.)

In cases where F(s) is known, $\bar{t}$ and $\sigma^2$ can be found directly in the Laplace domain. The method is based on the following set of operations:

| | Time Domain | Laplace Domain |
|---|---|---|
| Definition of Transform | f(t) | F(s) |
| Multiplication by t | tf(t) | -dF/ds |
| Integration | $\int_0^t tf(t)dt$ | $-\frac{1}{s}\frac{dF}{ds}$ |
| Final Value | $\int_0^\infty tf(t)dt$ | $\lim_{s\to 0}\frac{dF}{ds}$ |

From this final result we obtain

$$\bar{t} = \lim_{s\to 0}\frac{dF}{ds} \tag{10.18}$$

In a similar fashion, $\sigma_t^2$ can be evaluated from F(s):

$$\sigma_t^2 = \lim_{s\to 0}\frac{d^2F}{ds^2} - (\bar{t})^2 \tag{10.19}$$

**EXAMPLE 10.2** Given $F(s) = \frac{\alpha}{s+\alpha}$ what is the value of $\alpha$?

**Solution:** Example 10.1 showed that $F(s) = \frac{\alpha}{s+\alpha}$ can represent a residence time density function. To find $\alpha$, we apply Equation 10.13.

$$\bar{t} = \int_0^\infty tf(t) = \int_0^\infty t\alpha e^{-\alpha t}dt = \frac{1}{\alpha}$$

from which we conclude that $\alpha = (\bar{t})^{-1}$. The governing ODE for a first order system is thus

$$\bar{t}\,\frac{da_{out}}{dt} + a_{out} = a_{in}$$

This ODE corresponds to the flow of a nonreactive tracer through a perfectly mixed stirred tank. It turns out, however, that the residence time density function

$$f(t) = \frac{1}{\bar{t}}\,e^{-t/\bar{t}} \qquad \text{(i)}$$

can apply to systems which are **not** perfectly mixed on a molecular scale. This is a topic of substantial interest to chemical engineers. It is discussed in specialized books on **mixing**.

## 10.3 RESIDENCE TIMES IN IDEAL SYSTEMS

We will discuss two idealized flow systems which have very different residence time distributions.

### 10.3.1 Piston Flow

Piston flow is an idealization of highly turbulent flow in a long tubular reactor. All particles are supposed to have the same residence time so that the density function is

$$f(t) = \delta(t - \bar{t}), \qquad F(s) = e^{-s\bar{t}} \qquad (10.20)$$

The corresponding washout function is

$$w(t) = 1 - U(t - \bar{t}), \qquad W(s) = (1 - e^{-s\bar{t}})/s \tag{10.21}$$

(refer to the shifting property in Table 4.2). The variance for piston flow is found from Equation 10.15:

$$\sigma_t^2 = \int_0^\infty t^2 \delta(t - \bar{t})dt - (\bar{t})^2 = 0 \tag{10.22}$$

The dimensionless variance is also zero. This situation represents the smallest possible amount of mixing between particles which entered the system at different times, namely no mixing at all. Piston flow is seen to be one extreme limit for the behavior of a real flow system.

### 10.3.2 Perfect Mixing

The residence time density function for a perfectly mixed stirred tank is

$$f(t) = (1/\bar{t})e^{-t/\bar{t}}, \qquad F(s) = \frac{1}{\bar{t}s + 1} \tag{10.23}$$

(see Example 10.2). The corresponding washout function is

$$w(t) = e^{-t/\bar{t}}, \qquad W(s) = \frac{\bar{t}}{\bar{t}s + 1} \tag{10.24}$$

The variance for a perfect mixer can be found from Equation 10.19

$$\sigma_t^2 = \lim_{s \to 0} \frac{d^2F}{ds^2} - (\bar{t})^2 = \lim_{s \to 0} \frac{2(\bar{t})^2}{(\bar{t}s + 2)^3} - (\bar{t})^2 = (\bar{t})^2 \tag{10.25}$$

The dimensionless variance is 1.0, and this is a reasonable upper limit for most flow systems. Systems with **bypassing** or with laminar flow of low diffusivity fluids may exhibit $\sigma^2 > 1$. Perfect mixing is an extreme case of molecular level mixing that happens to give $\sigma^2 = 1$, but there are flow situations other than perfect mixing which will also give $\sigma^2 = 1$.

## 10.4 RESIDENCE TIMES IN COMPLEX FLOW SYSTEMS

Chapter 9 showed how to calculate transfer functions for composite systems. We will now use the versions of the summation and splitting rules which are applicable to flow systems. The transfer function for each subsystem will be the Laplace domain representation of the residence time distribution in that subsystem. The symbol $t = V/Q$ will be used for the composite system. The mean residence time in the jth subsystem is

$$t_j = \frac{V_j}{Q_j} \tag{10.26}$$

where $V_j$ is the volume of the jth subsystem and $Q_j$ is the flow rate through it. To avoid confusion, it is probably best to write the subsystem transfer functions using $V_j$ and $Q_j$ rather than $t_j$. In general, subsystem residence times are not additive (subsystems in series are an exception). To find t for the composite system, sum the $V_j$ to obtain the total system volume and then divide by the overall flow rate, Q.

With the above points in mind, we will find $F_c(s)$ for a number of example systems of chemical engineering importance.

**EXAMPLE 10.3** Find $F_c(s)$ and $\sigma^2$ for a piston flow reactor in series with a perfect mixer.

**Solution:** The system is

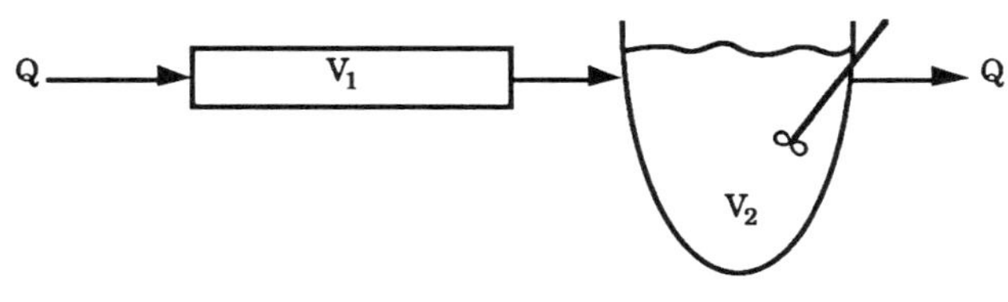


In block diagram form

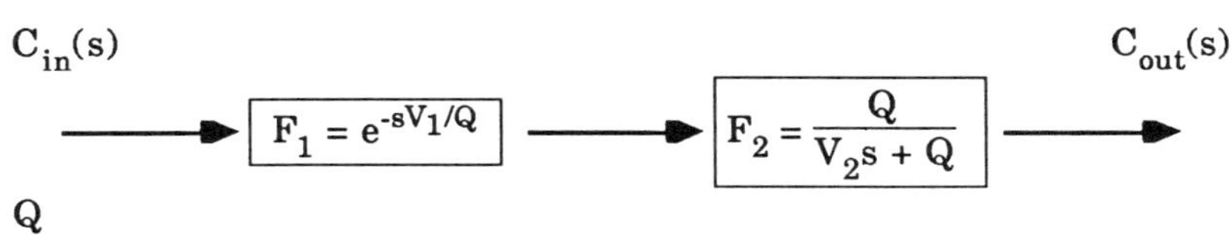


The overall transfer function is

$$F_c = F_1F_2 = \frac{Qe^{-sV_1/Q}}{V_2s + Q}$$

To find $\sigma^2$ we use Equation 10.19:

$$\frac{d^2F_c}{ds^2} = [\frac{V_1^2}{(V_2s+Q)Q} + \frac{2V_1V_2}{(V_2s+Q)^2} + \frac{2V_2^2Q}{(V_2s+Q)^3}]\ e^{-sV_1/Q}$$

$$\sigma_t^2 = \lim_{s \to 0} \frac{d^2F_c}{ds^2} - (\bar{t})^2 = \frac{V_1^2}{Q^2} + \frac{2V_1V_2}{Q_2} + \frac{2V_2^2}{Q_2} - (\bar{t})^2$$

where $\bar{t} = (V_1 + V_2)/Q$. Algebra gives

$$\sigma_t^2 = \frac{V_2^2}{Q^2}$$

For $\sigma^2$ we obtain

$$\sigma^2 = (\frac{V_2}{V_1+V_2})^2 = (\frac{V_2}{V})^2$$

Piston flow and perfectly mixed reactors in series can be used to model intermediate levels of mixing, $0 < \sigma^2 < 1$. If $V_2 = 0$, $\sigma^2 = 0$ and if $V_1 = 0$, $\sigma^2 = 1$.

---

**EXAMPLE 10.4** Find $F_c(s)$ and $\sigma^2$ for N equal volume stirred tanks in series.

**Solution:** The transfer function for a single tank is

$$F_j = \frac{Q}{V_js + Q} = \frac{Q}{Vs/N + Q} = \frac{1}{\bar{t}s/N + 1}$$

The composite transfer function is

$$F_c(s) = \prod_{j=1}^{N} F_j = \frac{1}{(\bar{t}s/N + 1)^N} = \frac{N^N}{(\bar{t}s + N)^N}$$

The variance is

$$\sigma_t^2 = \lim_{s \to 0} \frac{(N+1)N^{N+1}(\bar{t})^2}{(\bar{t}s + N)^{N+2}} - \bar{t}^2$$

and

$$\sigma^2 = 1/N.$$

---

**EXAMPLE 10.5** Determine $F_c(s)$ and $\sigma^2$ for a stirred tank with bypassing.

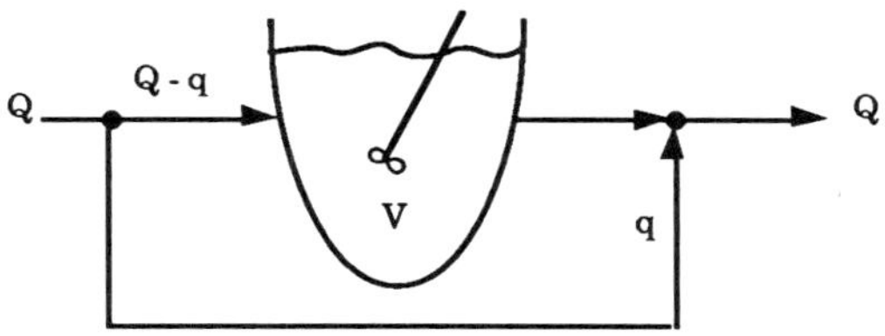


**Solution:** The block diagram can be written as

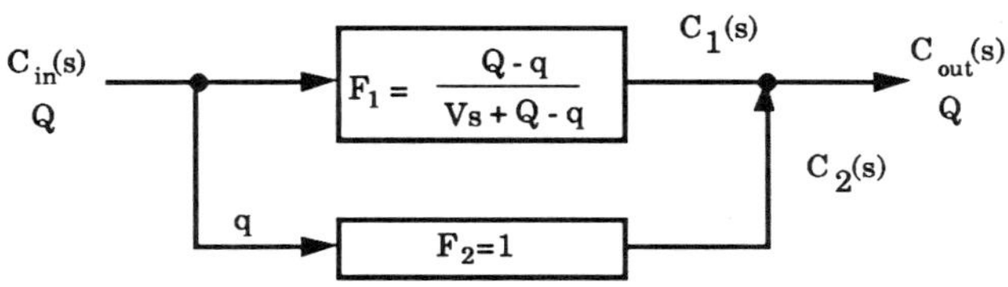


Detail about the addition point is

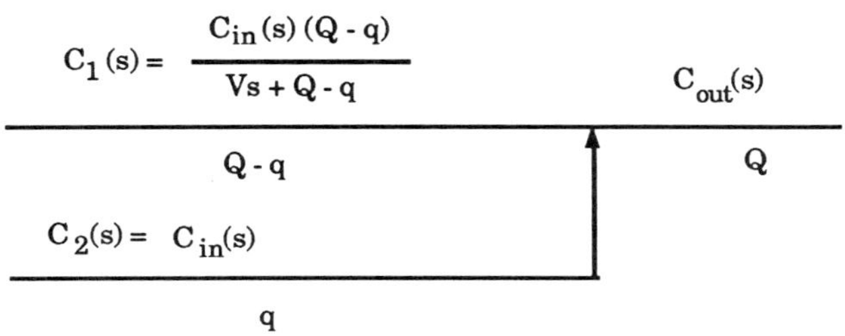


Applying the summation rules for a flow system addition point gives

$$C_{out}(s) = \frac{(Q-q)C_1(s)+qC_2(s)}{Q} = \frac{(Q-q)^2}{(Vs + Q - q)\, Q} + q/Q$$

and

$$\frac{C_{out}}{C_{in}} = F_c(s) = \frac{qVs + Q(Q\text{-}q)}{Q(Vs + Q - q)} = \frac{b\bar{t}s + 1 - b}{\bar{t}s + 1 - b}$$

where $\bar{t}$ = V/Q and b = q/Q is the **fraction bypassed**. The variance is found in the usual manner. For $\sigma^2$ we obtain

$$\sigma^2 = \frac{1 + b}{1 - b}$$

Bypassing gives $\sigma^2 > 1$. It is rarely desirable.

---

**EXAMPLE 10.6** Determine $F_c$ and $\sigma^2$ for a perfect mixer with instantaneous recycle.

**Solution:** The block diagram is now

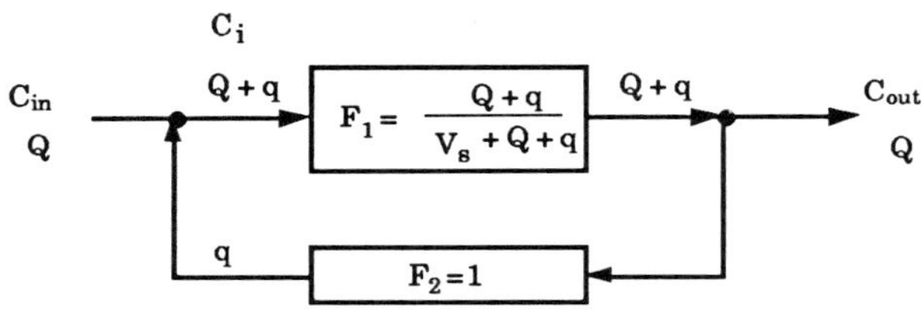


Detail about the summation point is

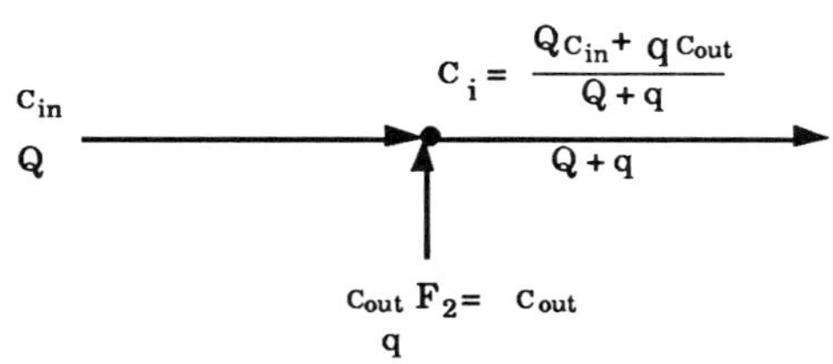


The concentration $C_i(s)$ is the net input to the stirred tank. The output is

$$C_{out} = C_i F_1 = \frac{QC_{in} + qC_{out}}{Vs + Q + q}$$

Solving for $C_{out}(s)$ gives

$$C_{out}(s) = \frac{QC_{in}}{Vs + Q}$$

or

$$\frac{C_{out}}{C_{in}} = F_C = \frac{Q}{Vs + Q} = \frac{1}{\bar{t}s + 1}$$

This is the same as for a perfect mixer **without** recycle. Instantaneous recycle has no effect on a perfect mixer since the contents of the vessel are already completely mixed.

---

The reader may note that the above examples were worked entirely in the Laplace domain. The results were obtained far more easily than they would have been in the time domain. The various results for $F_c(s)$ can be inverted to obtain $f_c(t)$. The inversions are apt to be messy, however.

## 10.5 USES OF RESIDENCE TIME DISTRIBUTIONS

Probably the most important use of residence time measurements is to form a qualitative understanding of flow patterns in otherwise unobservable equipment. Mere knowledge of the residence time distribution is seldom adequate to model a process, but this knowledge can provide considerable insight toward the development of a comprehensive model.

There are two important situations where the residence time distribution does provide enough information to predict system performance. These are the dampening of concentration fluctuations and the reaction of a compound according to first order, isothermal, homogeneous kinetics. They can be combined into a single equation which is a convolution integral in the time domain

$$a_{out}(t) = \int_0^t a_{in}(t-\tau)f(\tau)e^{-k\tau}\, d\tau \tag{10.27}$$

where k is the rate constant for the reaction.

The Laplace domain counterpart of Equation 10.27 is

$$A_{out}(s) = A_{in}(s)\, F\,(s+k) \tag{10.28}$$

These results give the outlet concentration of a reactive tracer having a time variable input, $a_{in}(t)$. The nonreactive case is found by setting $k = 0$.

**EXAMPLE 10.7** Find the steady state outlet concentration from an isothermal, homogeneous reactor having an arbitrary residence time density function f(t). Assume the reaction is first order and that the inlet concentration is constant. Apply your result to a single stirred tank, to N equal volume tanks in series, and to a piston flow reactor.

**Solution:** This is a steady state problem, but the answer can be found using the transient analysis inherent in Equation 10.28. Set $A_{in}(s) = a_{in}/s$ and apply the final value theorem.

$$a_{out}(\infty) = \lim_{s \to 0} sA_{in}(s)F(s+k) = a_{in} \lim_{s \to 0} F(s+k) = a_{in}F(k)$$

The result can also be found from Equation 10.27 by taking the limit as $t \to \infty$:

$$a_{out}(\infty) = a_{in} \int_0^{\infty} f(\tau)e^{-k\tau}\, d\tau = a_{in}F(k)$$

For perfect mixing, $F(s) = 1/(\bar{t}s + 1)$. Thus,

$$\frac{a_{out}}{a_{in}} = F(k) = \frac{1}{1 + k\bar{t}} \qquad \text{(i)}$$

For N tanks in series, we use a result from Example 10.4: $F(s) = N^N/(ts + N)^N$. Thus,

$$\frac{a_{out}}{a_{in}} = \frac{N^N}{(N + k\bar{t})^N} \qquad \text{(ii)}$$

For piston flow, $F(s) = e^{-s\bar{t}}$. Thus,

$$\frac{a_{out}}{a_{in}} = e^{-k\bar{t}} \qquad \text{(iii)}$$

Equation ii clearly reduces to Equation i if N = 1. We leave it to the mathematically inclined reader to show that Equation ii reduces to Equation iii in the limit of $N \to \infty$.

---

**EXAMPLE 10.8** A stirred tank is commonly used to damp out concentration fluctuations. Suppose the input stream of an inert tracer varies as

$$c_{in}(t) = c_o\,(1 + \alpha \sin\omega t), \quad 0 < \alpha < 1$$

Find the amplitude of the outlet concentration fluctuation.

**Solution:** For an inert tracer, k = 0 and Equation 10.28 gives

$$C_{in}(s) = C_{in}(s)\ F(s) = \left(\frac{c_o}{s} + \frac{c_o \alpha \omega}{s^2 + \omega^2}\right) \frac{1}{(s\bar{t} + 1)}$$

The periodic portion of the outlet response corresponds to the term containing $(s^2 + \omega^2)$, and we expand this term as a partial fraction

$$\frac{c_o \alpha \omega}{(s^2 + \omega^2)(s\bar{t} + 1)} = \frac{K_o + K_1 s}{s^2 + \omega^2} + \frac{K_2}{(s\bar{t} + 1)}$$

The $K_o + K_1 s$ term inverts to a phase-shifted sine wave with amplitude $K_o$. See Example 6.3. Thus, the outlet amplitude is

$$K_o = \frac{c_o \alpha}{1 + (\omega \bar{t})^2}$$

The magnitude of the sinusoidal ripple is decreased by the factor $1 + (\omega\bar{t})^2$. Note that dampening performance is governed by the dimensionless number $\omega t$.

---

Residence time measurements can be used to verify a flow model. Chapter 1 introduced the three components of systems analysis: modeling, solution, and verification. Most of this book is devoted to modeling and solution techniques since verification tends to be system specific. The stimulus-response methods of Section 10.1 are fairly general, but the type of stimulus obviously depends on the system. The electrical systems of Chapter 6 are stimulated by voltage inputs while residence time measurements using inert tracers are suitable for flow systems.

Suppose the residence time distribution of a process has been measured. How can this information be used to verify a model? Recognize that the residence time distribution will only verify the flow aspects of the model. An overall process model may include a factor such as temperature which is not measurable using inert tracers. All that can be expected is for the model prediction to fit the experimental measurements reasonably well. There are many possible ways to achieve this. You could, for example, use the method of moments described in Section 10.2 to force the model to have the same values of t and $\sigma^2$ as the real system. However, a method known as **time domain least squares analysis** is usually better for fitting the experimental response of a dynamic system. The procedure can be divided into the following steps:

1. Stimulate the system and measure its dynamic response, $[c_{out}(t)]_{system}$.

2. Mathematically stimulate the model to determine its response, $[c_{out}(t)]_{model}$.

3. Calculate the sum of the squared differences between the measurements and predictions

$$S = \sum_{\text{All data points}} \{[c_{out}(t)]_{system} - [c_{out}(t)]_{model}\}^2 \tag{10.29}$$

4. Adjust the parameters in the model to minimize the value of S.

**EXAMPLE 10.9** Suppose the residence time washout function has been measured as

| t(min) | w(t) |
|---|---|
| 0 | 1.00 |
| 15 | .79 |
| 30 | .60 |
| 45 | .45 |
| 60 | .38 |
| 75 | .27 |
| 90 | .21 |
| 105 | .18 |
| 120 | .12 |
| 150 | .07 |
| 180 | .05 |

Model the system as a piston flow reactor in series with a perfect mixer. Use the method of time domain least squares.

**Solution:** We must first find the model prediction for $c_{out}(t) = w(t)$. Example 10.3 gave

$$F_c = \frac{Qe^{-sV_1/Q}}{V_2 s + Q} = \frac{e^{-st_1}}{t_2 s + 1}$$

where $t_1 = V_1/Q$ and $t_2 = V_2/Q$ for this model system. Use Equation 10.12 and invert to find

$$\begin{aligned} w(t) &= 1, \ 0 < t < t_1 \\ &= \exp - (t - t_1)/t_2, \ t > t_2 \end{aligned}$$

This result completes step 2 in the procedure.

Steps 3 and 4 require a computer program. A BASIC program follows

```
        DIM T(11), W(11)
        DATA 0,15,30,45,60,75,90,105,120,150,180
        DATA 1.00, .79, .60, .45, .38, .27, .21, .18, .12, .07, .05
        FOR I = 1 to 11
        READ T(I)
        NEXT I
        FOR I = 1 to 11
        READ W(I)
        NEXT I

1       INPUT T1, T2
        S=0
        FOR I = 1 to 11
        IF T(I)<T1 THEN WMODEL=1 ELSE WMODEL=EXP(-(T(I)-T1)/T2)

        S = S + (W(I) - WMODEL)^2
        NEXT I
        PRINT S
        GO TO 1
```

A trial and error search is conducted for the combination of T1 = $t_1$ and T2 = $t_2$ that minimizes S. Results are $t_1$ = 1.00 minutes and $t_2$ = 57.66 minutes. Thus, according to the model, the system consists of a stirred tank reactor having nearly a 1 hour residence time preceded (or followed) by a small piston flow reactor having a residence time of 1 minute.

---

The above example used a simple technique, trial and error, for finding those values of the model parameters that minimized S. More elegant methods are available in computer packages for **nonlinear regression analysis**. Some of them are even user friendly. For a more complete discussion of residence time distributions and of the analysis of residence time data, see Nauman and Buffham, **Mixing in Continuous Flow Systems**, Wiley, New York, 1983.

## PROBLEMS

10.1 An experimentally obtained washout function is approximated by

$$w(t) = 1, \qquad 0 < t < 3$$

$$w(t) = e^{-2(t-3)}, \qquad t > 3$$

where t is in seconds. Find $\bar{t}$, $\sigma_t^2$, and $\sigma^2$.

10.2 Two equal volume stirred tanks are connected in parallel. Find $F_c(s)$ and $\sigma^2$ for this situation.

10.3 Find $F_c$ for the following flow system.

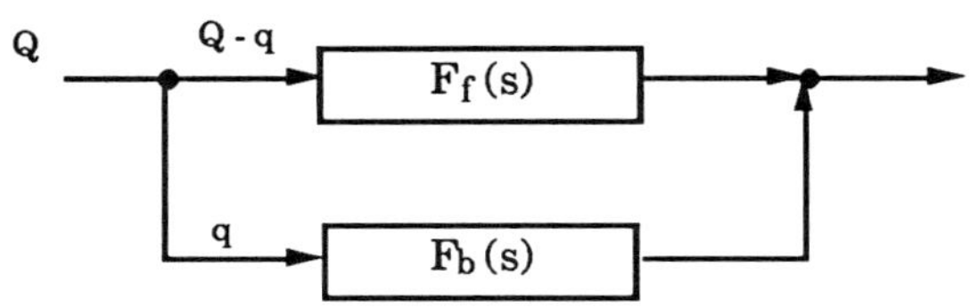


10.4 What physical system could give the following density function?

$$F(s) = \frac{1}{(s+1)(2s+1)}$$

Are there other possibilities than the one you chose?

10.5 Suppose an initially relaxed system is disturbed by a step change, $a_0U(t)$, of a reactive tracer having first order rate constant k. Find $a_{out}(t)$. How is this function related to w(t) for the case of k = 0?

10.6 Find the steady state reaction yield for the bypassing reactor in Example 10.5 given that the reaction is first order, isothermal, and homogeneous.

10.7 Repeat Example 10.8 for the case of a piston flow reactor.

# CHAPTER 11

# MATRIX DIFFERENTIAL EQUATIONS

The foregoing chapters of this text have shown that many engineering systems can be modeled using sets of ordinary differential equations. When these sets of ODEs are linear with constant coefficients, they can be solved by algebraic manipulations in the Laplace domain. The current chapter generalizes this approach using matrix algebra. The result will be an elegant, general solution which can be written as a single line of type. (Practical evaluation of this solution requires the ordinary and sometimes messy algebra used in the previous chapters of this book, but that is another story.)

## 1.1 MATRIX FORMULATION

Consider the general system illustrated below:

| **Input** | | **System** | | **Output** |
|---|---|---|---|---|
| Input Variables | → | Internal or State Variables | → | Output Variables |
| $u_1(t)$ | → | $x_1(t)$ | → | $v_1(t)$ |
| $u_2(t)$ | → | $x_2(t)$ | → | $v_1(t)$ |
| $\vdots$ | | $\vdots$ | | $\vdots$ |
| $u_L(t)$ | → | $x_M(t)$ | → | $v_N(t)$ |

The system is a multiple-input, multiple-output device. There are L inputs, N outputs, and M internal or **state variables** which together define the state of the system at any time. These various variables need not be the same physical quantities or have the same units. In a chemical system the inputs might be the flow rates of pure components plus energy. The state variables could be various chemical concentrations plus temperature and pressure (additional state variables are needed for multiphase systems). The output variables could be the product concentrations. For the isothermal, homogeneous stirred tank reactors treated in Chapters 2 and 10, the state variables and the output variables are identical. This is a special case that frequently arises in electrical engineering systems as well.

The collections of input, state, and output variables can be regarded as column vectors or matrices:

| **Input** | **System** | **Output** |
|---|---|---|
| $\mathbf{u}(t) = \begin{pmatrix} u_1(t) \\ u_2(t) \\ \vdots \\ u_L(t) \end{pmatrix}$, | $\mathbf{x}(t) = \begin{pmatrix} x_1(t) \\ x_2(t) \\ \vdots \\ x_M(t) \end{pmatrix}$, | $\mathbf{v}(t) = \begin{pmatrix} v_1(t) \\ v_2(t) \\ \vdots \\ v_N(t) \end{pmatrix}$ |
| (L x 1) | (M x 1) | (N x 1) |

(11.1)

These are called the **input vector, state vector,** and **output vector,** respectively. Their dimensions L, M, and N can all be different. We suppose the system is governed by a set of linear, first order ODEs:

$$\begin{aligned} \frac{dx_1}{dt} &= a_{11}x_1 + a_{12}x_2 + \dots + a_{1M}x_M + b_{11}u_1 + b_{12} + \dots + b_{1L}u_L \\ \frac{dx_2}{dt} &= a_{21}x_1 + a_{22}x_2 + \dots + a_{2M}x_M + b_{21}u_1 + b_{22}u_2 + \dots + b_{2L}u_L \\ &\vdots \\ \frac{dx_M}{dt} &= a_{M1}x_1 + a_{M2}x_2 + \dots + a_{MM}x_M + b_{M1}u_1 + b_{M2}u_2 + \dots + b_{ML}u_L \end{aligned} \tag{11.2}$$

There are M of these ODEs, one for each state variable. Each ODE is forced by a linear combination of the $u_l$.

$$f_m = \sum_{l=1}^{L} b_{ml}\, u_l \tag{11.3}$$

where m = 1,2, ..., M. Thus, as in earlier chapters, there is one forcing function per ODE and per state variable. The (Mx1) matrix of forcing functions is the **forcing vector, Bu**. Note that **u** is an (Lx1) matrix.

Equation 11.2 can be written in matrix notation as

$$\underset{(Mx1)}{\frac{d\mathbf{x}}{dt}} = \underset{(MxM)}{\mathbf{A}} \; \underset{(Mx1)}{\mathbf{x}} + \underset{(MxL)}{\mathbf{B}} \; \underset{(Lx1)}{\mathbf{u}} \tag{11.4}$$

We have indicated the size of the various matrices to show that the rules of matrix multiplication and addition are satisfied. **A** and **B** are matrices of coefficients. They are obtained directly from Equation 11.2

$$\mathbf{A} = \begin{pmatrix} a_{11} & a_{12} & \cdots & a_{1M} \\ a_{21} & a_{22} & \cdots & a_{2M} \\ \vdots & \vdots & & \vdots \\ a_{M1} & a_{M2} & \cdots & a_{MM} \end{pmatrix} \tag{11.5}$$

$$\mathbf{B} = \begin{pmatrix} b_{11} & b_{12} & \cdots & b_{1L} \\ b_{21} & b_{22} & \cdots & b_{2L} \\ \vdots & \vdots & & \vdots \\ b_{M1} & b_{M2} & \cdots & b_{ML} \end{pmatrix} \tag{11.6}$$

Equation 11.4 is a **matrix differential equation**. It is equivalent to Equation 11.2 but is written in more compact notation. Note that operations like differentiation and Laplace transformation are done element by element:

$$\frac{d\mathbf{x}}{dt} = \begin{pmatrix} \frac{dx_1}{dt} \\ \frac{dx_2}{dt} \\ \vdots \\ \frac{dx_M}{dt} \end{pmatrix} \quad \text{and } \pounds[\mathbf{x}(t)] = \mathbf{X}(s) = \begin{pmatrix} X_1(s) \\ X_2(s) \\ \vdots \\ X_M(s) \end{pmatrix} \tag{11.7}$$

The solution to Equation 11.4 will be the (Mx1) vector of state variables, **x**. The solution is subject to a vector of initial conditions, **x**(0). The next section will solve Equation 11.4 using Laplace transformation techniques. However, a solution for **x**(t) does not directly give the output response of the system. We still need a relationship between **x**(t) and **v**(t). We use a linear form

$$\underset{(Nx1)}{\mathbf{v}(t)} = \underset{(NxM)\,(Mx1)}{\mathbf{C}\,\mathbf{x}(t)} + \underset{(NxL)\,(LX1)}{\mathbf{D}\mathbf{u}(t)} \tag{11.8}$$

where **C** is an (NxM) matrix of coefficients relating the outputs to the state variables and **D** is an (NxL) matrix of coefficients relating the output to the input variables. Normally, **v**(t) depends only on **x**(t) so that **D** = **0** is the **null**

**matrix**. However, the more general form of Equation 11.8 is useful for some applications.

**EXAMPLE 11.1** Write a model in matrix form for the following flow system:

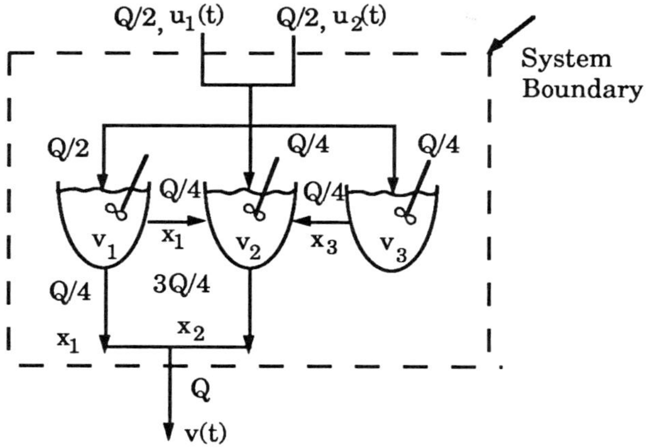


Here, $u_1$ and $u_2$ represent independently variable input concentrations of an inert tracer. The state variables $x_1$, $x_2$, and $x_3$ denote internal concentrations of the same tracer, and v is its outlet concentration.

**Solution:** We have L = 2, M = 3, and N = 1. A material balance about the upper mixing point gives

$$(x_1)in = (x_2)in = (x_3)in = \frac{u_1(t)}{2} + \frac{u_2(t)}{2}$$

The dynamic equations for the various stirred tanks are

| Accumulation | = | input | - | output |
|---|---|---|---|---|
| $V_1 \frac{dx_1}{dt}$ | = | $\frac{Q(x_1)_{in}}{2}$ | - | $\frac{Q}{2} x_1$ |
| $V_2 \frac{dx_2}{dt}$ | = | $\frac{Q(x_2)_{in}}{4} + \frac{Q}{4} x_1 + \frac{Q}{4} x_3$ | - | $\frac{3Qx_2}{4}$ |
| $V_3 \frac{dx_3}{dt}$ | = | $\frac{Q(x_3)_{in}}{4}$ | - | $\frac{Q}{4} x_3$ |

This can be written in explicit matrix form as

$$\frac{d}{dx}\begin{pmatrix} x_1 \\ x_2 \\ x_3 \end{pmatrix} = \begin{pmatrix} \frac{-Q}{2V_1} & 0 & 0 \\ \frac{Q}{4V_2} & \frac{-3Q}{4V_2} & \frac{Q}{4V_2} \\ 0 & 0 & \frac{-Q}{4V_3} \end{pmatrix} \begin{pmatrix} x_1 \\ x_2 \\ x_3 \end{pmatrix} + \begin{pmatrix} \frac{Q}{4V_1} & \frac{Q}{4V_1} \\ \frac{Q}{4V_2} & \frac{Q}{4V_2} \\ \frac{Q}{4V_3} & \frac{Q}{4V_3} \end{pmatrix} \begin{pmatrix} u_1 \\ u_2 \end{pmatrix}$$

(3x1) (3x3) (3x1) (3x2) (2x1)

A material balance on the lower mixing point gives

$$v(t) = \frac{x_1}{4} + \frac{3x_2}{4}$$

The matrix form of the output equation is

$$v = \left(\frac{1}{4} \quad 3/4 \quad 0\right) \begin{pmatrix} x_1 \\ x_2 \\ x_3 \end{pmatrix}$$

(1x1) (1x3) (3x1)

---

## 11.2 A GENERAL SOLUTION

Suppose the coefficient matrices **A** and **B** in Equation 11.4 are constant, independent of time. In this case, Equation 11.4 represents a set of linear, first order ODEs with constant coefficients. We can take the Laplace transform to obtain

$$s\mathbf{X}(s) - \mathbf{x}(0) = \mathbf{AX}(s) + \mathbf{BU}(s) \tag{11.9}$$

The problem at this point is to solve for $\mathbf{X}(s)$ and then to invert to obtain $\mathbf{x}(t)$. Equation 11.9 can be rearranged to give

$$s\mathbf{X} - \mathbf{AX} = \mathbf{BU} + \mathbf{x}(0) \tag{11.10}$$

The left-hand side of Equation 11.10 may be factored in matrix form to give

$$(s\mathbf{I} - \mathbf{A})\mathbf{X} = \mathbf{BU} + \mathbf{x}(0)$$

where **I** is the (MxM) identity matrix, consisting of ones on the diagonal and zeros elsewhere. The composite matrix (s**I** - **A**) is (MxM) square and thus can have an inverse, $(s\mathbf{I} - \mathbf{A})^{-1}$. Suppose the inverse exists. Premultiply by the inverse to obtain

$$\mathbf{X}(s) = (s\mathbf{I} - \mathbf{A})^{-1}\,[\mathbf{BU} + \mathbf{x}(0)] \tag{11.11}$$

This is the desired solution for **X**(s). The time domain solution is given by

$$\mathbf{x}(t) = \pounds^{-1}[\mathbf{X}(s)] = \pounds^{-1}[(s\mathbf{I} - \mathbf{A})^{-1}\,[\mathbf{BU} - \mathbf{x}(0)]] \tag{11.12}$$

where the inversion, like the forward transformation in Equation 11.7, is performed term by term.

**EXAMPLE 11.2** Use matrix methods to solve the following set of equations.

$$\frac{dx_1}{dt} = -2x_1 - x_2 + \delta(t)$$

$$\frac{dx_2}{dt} = -x_1 - 2x_2 + \delta(t)$$

The initial conditions are $x_1(0) = 0$, $x_2(0) = -1$.

**Solution:** The set can be written as

$$\frac{d}{dt}\begin{pmatrix} x_1 \\ x_2 \end{pmatrix} = \begin{pmatrix} -2 & -1 \\ -1 & -2 \end{pmatrix}\begin{pmatrix} x_1 \\ x_2 \end{pmatrix} + \begin{pmatrix} 1 \\ 1 \end{pmatrix}\delta(t)$$

which has the form of Equation 11.4. Thus,

$$\mathbf{A} = \begin{pmatrix} -2 & -1 \\ -1 & -2 \end{pmatrix}$$

and

$$s\mathbf{I} - \mathbf{A} = \begin{pmatrix} s+2 & 1 \\ 1 & s+2 \end{pmatrix}$$

Use your favorite method to invert this matrix. Whatever method you use should give

$$(s\mathbf{I} - \mathbf{A})^{-1} = \begin{pmatrix} \dfrac{s+2}{(s+1)(s+3)} & \dfrac{-1}{(s+1)(s+3)} \\ \dfrac{-1}{(s+1)(s+3)} & \dfrac{s+2}{(s+1)(s+3)} \end{pmatrix}$$

You can confirm this by multiplying

$$(s\mathbf{I} - \mathbf{A})^{-1}\,(s\mathbf{I} - \mathbf{A}) = \mathbf{I}$$

The matrix **BU** + **x**(0) becomes

$$\begin{pmatrix}1\\1\end{pmatrix}(1) + \begin{pmatrix}0\\-1\end{pmatrix} = \begin{pmatrix}1\\0\end{pmatrix}$$

Thus, Equation 11.11 gives

$$\begin{pmatrix}X_1\\X_2\end{pmatrix} = \begin{pmatrix}\dfrac{s+2}{(s+1)(s+3)} & \dfrac{-1}{(s+1)(s+3)}\\ \dfrac{-1}{(s+1)(s+3)} & \dfrac{s+2}{(s+1)(s+3)}\end{pmatrix}\begin{pmatrix}1\\0\end{pmatrix}$$

$$= \begin{pmatrix}\dfrac{s+2}{(s+1)(s+3)}\\ \dfrac{-1}{(s+1)(s+3)}\end{pmatrix} = \begin{pmatrix}\dfrac{1/2}{s+1} + \dfrac{1/2}{s+3}\\ \dfrac{-1/2}{s+1} + \dfrac{1/2}{s+3}\end{pmatrix}$$

To obtain the last matrix we used the usual approach of partial function expansion. This result is inverted term by term to give

$$\begin{pmatrix}x_1(t)\\x_2(t)\end{pmatrix} = \begin{pmatrix}\dfrac{e^{-t} + e^{-3t}}{2}\\ \dfrac{-e^{-t} + e^{-3t}}{2}\end{pmatrix}$$

---

**EXAMPLE 11.3** Revisit Example 11.2 using the ordinary algebra of the previous chapters.

**Solution:** Laplace transform the governing equations to obtain

$$sX_1 - 0 = -2X_1 - X_2 + 1$$

$$sX_2 + 1 = -X_1 - 2X_2 + 1$$

We need to solve these for $X_1$ and $X_2$. Rearrange them in the form

$$(s+2)X_1 + X_2 = 1$$

$$X_1 + (s+2)X_2 = 0$$

This is the set to be solved, and algebra will do it. Stop, however, and note that this set can be written as

$$\begin{pmatrix} s+2 & 1 \\ 1 & s+2 \end{pmatrix} \begin{pmatrix} X_1 \\ X_2 \end{pmatrix} = \begin{pmatrix} 1 \\ 0 \end{pmatrix}$$

The algebraic solution has the same effect as inverting the matrix of coefficients:

$$\begin{pmatrix} X_1 \\ X_2 \end{pmatrix} = \begin{pmatrix} s+2 & 1 \\ 1 & s+2 \end{pmatrix}^{-1} \begin{pmatrix} 1 \\ 0 \end{pmatrix}$$

This is exactly the same result as obtained in Example 11.2

---

**EXAMPLE 11.4** Find the **zero input response** for a set of linear, first order ODEs with constant coefficients.

**Solution:** This is a special case of the general solution, Equations 11.11 and 11.12. For this case,

$$\mathbf{u}(t) = \mathbf{O}, \qquad \mathbf{U}(s) = \mathbf{O}$$

so that

$$\mathbf{X}(s) = (s\mathbf{I} - \mathbf{A})^{-1}\, \mathbf{x}(0) \qquad \text{(i)}$$

and

$$\mathbf{x}(t) = \{\pounds^{-1}[(s\mathbf{I} - \mathbf{A})^{-1}]\}\, \mathbf{x}(0) \qquad \text{(ii)}$$

The matrix $(s\mathbf{I} - \mathbf{A})^{-1}$ is common to all solutions of the governing ODEs. It is sometimes called the **resolvent matrix**. Its inverse Laplace transformation, which appears in Equation ii, is called the **state-transition matrix.**

---

## 11.3 TRANSFER FUNCTIONS

Chapter 9 introduced the idea of transfer functions for single-input, single-output devices. In that treatment, the state and output variables were regarded as the same physical quantities. Typically, the input variable was also the same physical quantity as the output variable since this led to transfer functions which conserved mass. The idea of **transfer function** will now be generalized to systems with multiple-inputs and multiple-outputs. In general, the input, state, and output voltages will be different physical quantities. Consider a coin operated vending machine. The inputs are a mixture of coins plus a manual selection. The state variables are electrical and mechanical signals. The output variables are returned change plus purchased goods. Conceptually, at least, the transfer functions defined below will handle such a diverse array of physical quantities.

We define the **transfer function matrix**, **F**(s), by the equation

$$\underset{(N\times 1)}{\mathbf{V}(s)} = \underset{(N\times L)}{\mathbf{F}(s)} \; \underset{(L\times 1)}{\mathbf{U}(s)} \qquad (11.13)$$

Division is not allowed in matrix algebra, so this definition is as close as we can come to stating that this transfer function is equal to the ratio, in the Laplace domain, of the inputs to the outputs.

Applying Laplace transformation to Equation 11.8 gives

$$\mathbf{CX} + \mathbf{DU} = \mathbf{V} = \mathbf{FU}$$

The general solution of Equation 11.11 is now used to eliminate **X**. However, we first restrict attention to initially relaxed systems with $\mathbf{x}(0) = \mathbf{0}$. Substituting for **X** gives

$$\mathbf{C}(s\mathbf{I} - \mathbf{A})^{-1}\mathbf{BU} + \mathbf{DU} = \mathbf{FU}$$

Note that **U** is a common factor. The **Laplace-domain transfer function matrix** is

$$\underset{(N\times L)}{\mathbf{F}(s)} = \underset{(N\times M)}{\mathbf{C}} \quad \underset{(M\times M)}{(s\mathbf{I} - \mathbf{A})^{-1}} \; \underset{(M\times L)}{\mathbf{B}} \quad \underset{(N\times L)}{+\mathbf{D}} \qquad (11.14)$$

The **time-domain, transfer-function matrix** is obtained using term by term Laplace inversion:

$$\mathbf{f}(t) = \mathcal{L}^{-1}[\mathbf{F}(s)] \qquad (11.15)$$

There is no fixed relationship between L, M, and N. To avoid playfully trivial cases, we suppose each of these integers is at least one. Systems with more outputs than inputs are allowed. Also, the outputs need not be linearly independent. For example, $v_1 = x_1$, $v_2 = x_2$, and $v_3 = x_1 + x_2$ are perfectly acceptable.

**EXAMPLE 11.5** Determine the transfer function and the output response for the following situation:

$$u = 6e^{-t}$$

$$\frac{dx}{dt} = -2x + u$$

$$v = 3x + 4u$$

**Solution:** This case has L = M = N = 1. The various matrices become

$$\mathbf{A} = (-2) \qquad \mathbf{B} = (1) \qquad \mathbf{C} = (3) \qquad \mathbf{D} = (4)$$

$$\mathbf{I} = (1) \qquad (s\mathbf{I} - \mathbf{A}) = (s+2) \qquad (s\mathbf{I} = \mathbf{A})^{-1} = (\frac{1}{s+2})$$

Equation 11.14 gives

$$\mathbf{F} = (\frac{3}{s+2} + 4)$$

and Equation 11.15 gives

$$\mathbf{f} = (3e^{-2t} + 4\delta(t))$$

The response of the system to $u(t) = 6e^{-t}$ is found using Equation 11.13.

$$\mathbf{V}(s) = \mathbf{F}(s)\,\mathbf{U}(s) = (\frac{3}{s+2} + 4)(\frac{6}{s+1}) = (\frac{18}{(s+2)(s+1)} + \frac{24}{s+1}) = (\frac{42}{s+1} - \frac{18}{s+2})$$

The time domain output is

$$\mathbf{v}(t) = (42e^{-t} - 18e^{-2t})$$

---

**EXAMPLE 11.6** Draw a block diagram for the system in Example 11.5.

**Solution:** We use the voltage form of block diagrams since the output contains gains, $\mathbf{C} = (3)$ and $\mathbf{D} = (4)$, which cannot be readily justified in a flow system.

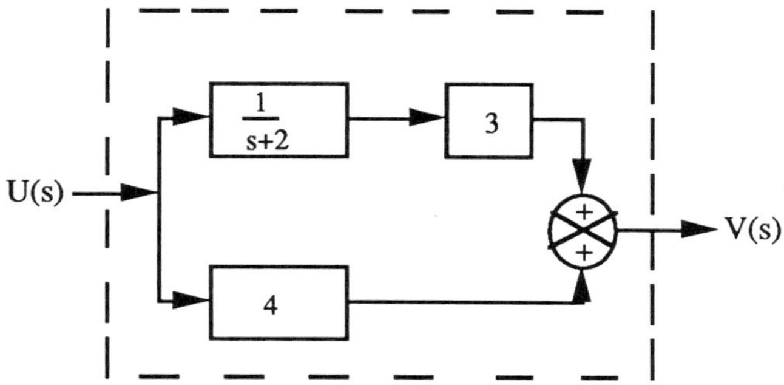


For a parallel system (Example 9.2),

$$F_c = F_1 + F_2 = \frac{3}{s+2} + 4$$

in agreement with Example 11.5.

---

**EXAMPLE 11.7** Determine the transfer function and draw a block diagram for the following system

$$\frac{dx_1}{dt} = -2x_1 + u_1$$

$$\frac{dx_2}{dt} = 4u_2$$

$$v = 3x_1 + x_2$$

**Solution:** We now have L = M = 2, N = 1. The various matrices become

$$\mathbf{A} = \begin{pmatrix} -2 & 0 \\ 0 & 0 \end{pmatrix} \quad \mathbf{B} = \begin{pmatrix} 1 & 0 \\ 0 & 4 \end{pmatrix} \quad \mathbf{C} = (3\ \ 1) \quad \mathbf{D} = \mathbf{0}$$

$$\mathbf{I} = \begin{pmatrix} 1 & 0 \\ 0 & 1 \end{pmatrix} \quad (s\mathbf{I} - \mathbf{A}) = \begin{pmatrix} s+2 & 0 \\ 0 & s \end{pmatrix} \quad (s\mathbf{I} - \mathbf{A})^{-1} = \begin{pmatrix} \frac{1}{s+2} & 0 \\ 0 & \frac{1}{s} \end{pmatrix}$$

Equation 11.14 gives

$$\mathbf{F} = (3\ \ 1) \begin{pmatrix} \frac{1}{s+2} & 0 \\ 0 & \frac{1}{s} \end{pmatrix} \begin{pmatrix} 1 & 0 \\ 0 & 4 \end{pmatrix} = \left( \frac{3}{s+2}\ \ \frac{4}{s} \right)$$

The output response is

$$\mathbf{V} = \mathbf{FU} = \left(\frac{3}{s+2} \;\; \frac{4}{s}\right)\begin{pmatrix} U_1 \\ U_2 \end{pmatrix} = \frac{3U_1(s)}{s+2} + \frac{4U_2(s)}{s}$$

The block diagram is

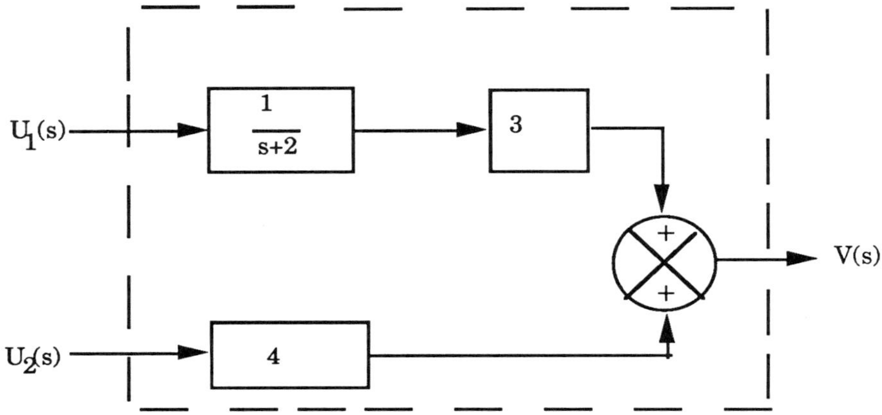


The reader will observe that this system behaves like that in Examples 11.5 and 11.6 for the special case of

$$U_2(s) = sU_1(s)\,, \qquad \frac{dU_2(t)}{dt} = u_1(t)$$

Another dual-input system which **always** behaves like that in examples 11.5 and 11.6 is

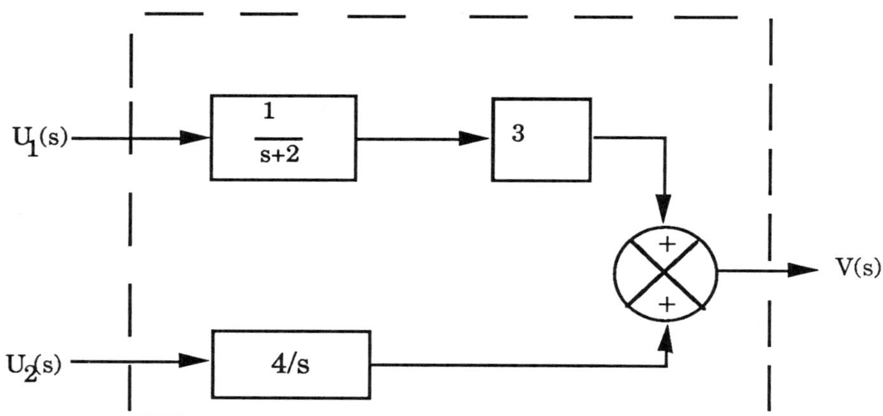

(In essence, we have just moved the voltage splitting point outside the system boundary.) In both these cases we have avoided the need for **D** by adding another input and another state variable. The allocation between the various variables is sometimes quite arbitrary.

---

**EXAMPLE 11.8** Consider the following second order system with negative feedback:

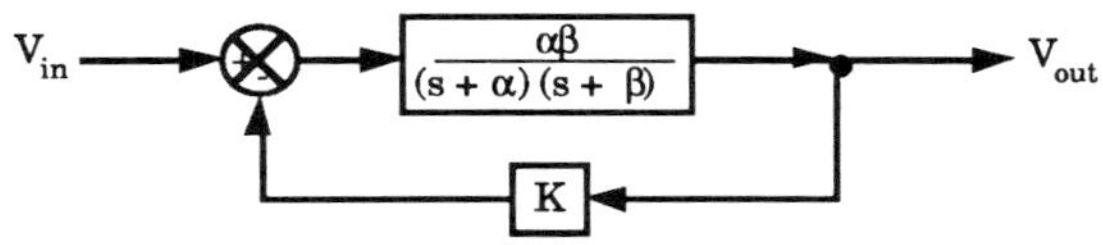


Determine the composite transfer function by treating this as a dual input system with $U_1 = U$ and $U_2 = -KV$.

**Solution:** The dual input system is

$$\begin{matrix} U & \rightarrow \\ -KV & \rightarrow \end{matrix} \quad \boxed{\frac{\alpha\beta}{(s+\alpha)(s+\beta)}} \rightarrow V$$

Apply Equation 11.13 with L = 2 and N = 1 to give

$$\underset{(1\text{x}1)}{(V)} = \underset{(1\text{x}2)}{\left(\frac{\alpha\beta}{(s+\alpha)(s+\beta)} \quad \frac{\alpha\beta}{(s+\alpha)(s+\beta)}\right)} \underset{(2\text{x}1)}{\begin{pmatrix} U \\ -KV \end{pmatrix}}$$

Performing the matrix multiplication gives

$$V = \frac{\alpha\beta U}{(s+\alpha)(s+\beta)} - \frac{K\alpha\beta V}{(s+\alpha)(s+\beta)}$$

Solve this for V/U to obtain

$$F_c = \frac{V}{U} = \frac{\alpha\beta}{(s+\alpha)(s+\beta) + K\alpha\beta}$$

This result is the same as that in Example 9.6.

---

**EXAMPLE 11.9** Use the matrix notation in Section 11.1 to describe the second order feedback system of Example 11.8.

**Solution:** The matrix formalism assumes that the system is governed by a set of first order ODEs. The second order system is governed by

$$\frac{d^2x}{dt^2} + (\alpha + \beta)\frac{dx}{dt} + \alpha\beta x = \alpha\beta u \qquad \text{(i)}$$

Here, $\alpha\beta u$ is the forcing function, with u having the same units as x.

The open-loop response is governed by Equation i with

$$u = u_{in}$$

The closed-loop response of the negative feedback system is governed by Equation i with

$$u = u_{in} - Kx$$

An equivalent set of first order ODEs is

$$\frac{dx}{dt} = y$$

$$\frac{dy}{dt} = -(\alpha+\beta)y - \alpha\beta(1+K)x + \alpha\beta u_{in}$$

The system is assumed initially relaxed in all cases. In explicit matrix notation the system becomes

$$\begin{pmatrix} \frac{dx}{dt} \\ \frac{dy}{dt} \end{pmatrix} = \begin{pmatrix} 0 & 1 \\ -\alpha\beta(1+K) & -\alpha-\beta \end{pmatrix} \begin{pmatrix} x \\ y \end{pmatrix} + \begin{pmatrix} 0 & 0 \\ 0 & \alpha\beta \end{pmatrix} \begin{pmatrix} 0 \\ u_{in} \end{pmatrix}$$

or $\frac{d\mathbf{x}}{dt} = \mathbf{Ax} + \mathbf{Bu}$

From this

$$s\mathbf{I} - \mathbf{A} = \begin{pmatrix} s & -1 \\ \alpha\beta(1+K) & s+\alpha+\beta \end{pmatrix}$$

and the resolvent matrix is

$$(s\mathbf{I} - \mathbf{A})^{-1} = \frac{\begin{pmatrix} s(s+\alpha+\beta) & +s \\ -s\alpha\beta(1+K) & s^2 \end{pmatrix}}{s[(s+\alpha)(s+\beta) + K\alpha\beta]}$$

To determine the transfer function matrix from Equation 11.14, we first need to find the matrices **C**, **B**, and **D**. Matrices **C** and **D** are defined by Equation 11.8. For the case at hand there is a single output, v = x. Thus, using Equation 11.8,

$$(v) = \mathbf{C}\begin{pmatrix} x \\ y \end{pmatrix} = (1 \quad 0)\begin{pmatrix} x \\ y \end{pmatrix}$$

and **D** = **0**. The matrix **B** has already been found as

$$\mathbf{B} = \begin{pmatrix} 0 & 0 \\ 0 & \alpha\beta \end{pmatrix}$$

Equation 11.14 gives

$$\mathbf{F}(s) = \mathbf{C}(s\mathbf{I}\text{-}\mathbf{A})^{-1}\,\mathbf{B} + \mathbf{D}$$

$$= (1 \quad 0)\ \frac{\begin{pmatrix} s(s+\alpha+\beta) & +s \\ -s\alpha\beta(1+K) & s^2 \end{pmatrix}}{s[(s+\alpha)(s+\beta) + K\alpha\beta]}\begin{pmatrix} 0 & 0 \\ 0 & \alpha\beta \end{pmatrix}$$

$$= \left(0 \quad \frac{\alpha\beta}{(s\alpha)\,(s+\beta) + K\alpha\beta}\right)$$

Finally, we apply Equation 11.13 to obtain

$$\mathbf{V} = \mathbf{F}\,\mathbf{u} = \left(0 \quad \frac{\alpha\beta}{(s+\alpha)\,(s+\beta) + K\alpha\beta}\right)\begin{pmatrix} 0 \\ u_{in} \end{pmatrix}$$

or

$$X(s) = \frac{\alpha\beta u_{in}}{(s+\alpha)\,(s+\beta) + K\alpha\beta}$$

in agreement with Examples 9.6 and 11.8.

---

## 11.4 CHARACTERISTIC EQUATIONS AND STABILITY

The MxM matrix

$$s\mathbf{I} - \mathbf{A} = \begin{pmatrix} s\text{-}a_{11} & -a_{12} & \cdots & -a_{1M} \\ -a_{21} & s\text{-}a_{22} & & \\ \vdots & & & \\ -a_{M1} & -a_{M2} & \cdots & s\text{-}a_{MM} \end{pmatrix} \qquad (11.16)$$

plays a key role in our treatment of matrix differential equation: we require its inverse if the set of equations is to be solved. It turns out that the inverse always exists, although it may be difficult to evaluate.

A square matrix has an inverse provided its determinant is not zero. Thus, we examine the equation

$$\det[(s\mathbf{I} - \mathbf{A})] = 0 \tag{11.17}$$

This equation yields a polynomial of degree M in s with a leading coefficient of one:

$$s^M + \beta_1 s^{M-1} + \beta_2 s^{M-2} + \ldots + \beta_{M-1} s + \beta_M = 0 \tag{11.18}$$

Equation 11.18 is known as the **characteristic equation** of the matrix. As a polynomial, it cannot be identically zero and can vanish only at a few isolated points, these being the M roots of Equation 11.18. This does not impair the invertibility of (s**I** - **A**) at all other values of s, so that an inverse exists almost everywhere in the complex plane. (Note that the roots of Equation 11.18 may occur in complex conjugates so that s must be regarded as a complex variable.)

The roots of Equation 11.18 are called the **characteristic values** or **eigenvalues** of the matrix **A**. They play a major role in matrix algebra. In the present application, the eigenvalues are identical to the **characteristic roots of the system**. They are the poles of the Laplace transform $(s\mathbf{I} - \mathbf{A})^{-1}$:

$$(s\mathbf{I} - A)^{-1} = \frac{\mathbf{N}(s)}{\det[(s\mathbf{I} - \mathbf{A})]} = \frac{\mathbf{N}(s)}{s^M + \beta_1 s^{M-1} + \ldots \beta_M}$$

$$= \frac{\mathbf{N}(s)}{(s-s_1)(s-s_2)\ldots(s-s_M)} \tag{11.19}$$

where **N**(s) is an MxM matrix, each term of which is a polynomial in s of degree $M^{-1}$ or lower. The matrix **N** is called the **adjoint matrix** and is usually denoted as adj[(s**I** - **A**)]. Laplace inversion of $(s\mathbf{I} - \mathbf{A})^{-1}$ can be performed term by term to obtain the **state-transition matrix** (see Example 11.4). A typical term of $(s\mathbf{I} - \mathbf{A})^{-1}$ has the form

$$\frac{N_{ij}(s)}{\det[(s\mathbf{I} - \mathbf{A})]} = \frac{N_{ij}(s)}{(s-s_1)(s-s_2)\ldots(s-s_M)} \tag{11.20}$$

and can be inverted using the methods of Chapter 4. The specifics of the inversion depend on **N**(s) = adj[(s**I** - **A**)]. For a higher order system, say $M > 3$, finding **N**(s) and performing the inversion can be quite difficult. However, a great deal of information can be found just from det[(s**I** - **A**)], which is somewhat easier to evaluate.

Suppose det[(s**I** - **A**)] is known and has been factored to give the roots $s_1$, $s_2$, ..., $s_M$. As always, the roots may be real or complex. Some roots may be repeated. If the real portion of **any** root is positive, the system is unstable and long time solutions will diverge to infinity. If the real portions of **all** roots are negative, the system is stable and long time solutions will approach a steady state. If there are complex roots, the system will oscillate. Strictly imaginary roots cause sustained oscillations. Complex roots will give divergent oscillations if the real parts are positive and damped oscillations if the real parts are negative. A single pole at s = 0 will allow attainment of a steady state, but multiple poles at s = 0 will not.

There are a few, typically trivial, exceptions to the above catalog of rules. Note also that we are considering stability only in the context of the open loop response. Unstable systems can be stabilized with feedback control.

**EXAMPLE 11.10** Does invertibility of (s**I** - **A**) imply invertibility of **A**?

**Solution:** No. Example 11.7 gave a counterexample with

$$\mathbf{A} = \begin{pmatrix} -2 & 0 \\ 0 & 0 \end{pmatrix}$$

where det**A** = 0 so that $A^{-1}$ does not exist. However, (s**I** - **A**) remains invertible even when **A** is the null matrix. For M = 2 and **A** = **0**:

$$(s\mathbf{I} - \mathbf{A}) = \begin{pmatrix} s & 0 \\ 0 & s \end{pmatrix}, \quad (s\mathbf{I} - \mathbf{A})^{-1} = \begin{pmatrix} \frac{1}{s} & 0 \\ 0 & \frac{1}{s} \end{pmatrix}$$

This example is one of the above-mentioned exceptions. There is a repeated pole at s = 0, but the zero-input response achieves a steady state.

The reason for this exception may be seen from Equation 11.20:

$$\frac{N_{11}}{\det[(s\mathbf{I} - \mathbf{A})])} = \frac{s}{s^2} = \frac{1}{s} = \frac{N_{22}}{\det[(\mathbf{I} - \mathbf{A})]}$$

so that one factor of s cancels between the numerator and the denominator.

---

**EXAMPLE 11.11** Determine the stability of the following initially relaxed system.

$$\frac{dx_1}{dt} = -x_1 + 2x_2 + u_1$$

$$\frac{dx_2}{dt} = 2x_1 - 2x_2 + u_2$$

**Solution:**

$$\mathbf{A} = \begin{pmatrix} -1 & 2 \\ 2 & -2 \end{pmatrix}$$

$$s\mathbf{I} - \mathbf{A} = \begin{pmatrix} s+1 & -2 \\ -2 & s+2 \end{pmatrix}$$

$$\det[s\mathbf{I} - \mathbf{A}] = (s+1)(s+2) - 4 = s^2 + 3s - 2$$

The roots of the characteristic equation are

$$s_1 = +0.56, \qquad s_2 = -3.56$$

Since a root is positive, the system is unstable. (If both $u_1(t) \equiv 0$ and $u_2(t) \equiv 0$, the outlet response will remain zero; but any nonzero input, no matter how small, will cause a divergent response.)

---

**EXAMPLE 11.12** Determine the stability of the following system.

$$\frac{d}{dt}\begin{pmatrix} x1 \\ x2 \\ x3 \end{pmatrix} = \begin{pmatrix} -1 & 2 & 0 \\ 2 & -2 & -3 \\ 0 & 2 & -1 \end{pmatrix}\begin{pmatrix} x_1 \\ x_2 \\ x_3 \end{pmatrix} + \begin{pmatrix} 1 & 0 & 0 \\ 0 & 1 & 0 \\ 0 & 0 & 1 \end{pmatrix}\begin{pmatrix} u_1 \\ u_2 \\ u_3 \end{pmatrix}$$

**Solution:** For the stability analysis being performed here, we need only be concerned with the roots of

$$\det[s\mathbf{I} - \mathbf{A})] = 0.$$

These are

$$\det\begin{pmatrix} s+1 & -2 & 0 \\ -2 & s+2 & +3 \\ 0 & -2 & s+1 \end{pmatrix} = (s+1)^2(s+2) - 4(s+1) + 6(s+1)$$

$$= (s+1)(s^2 + 3s + 4) = 0$$

The roots are

$$s_1 = -1\,, \qquad s_2 = -1.5 + 1.32i\,, \qquad s_2 = -1.5 - 1.32i$$

The real parts of all the roots are negative so the system will be stable. The pair of complex conjugates with negative real parts will lead to a damped oscillation.

---

Manual evaluation of determinates becomes tedious for a system of order greater than three. Also, factoring the characteristic equation may be difficult for orders greater than two (we cheated on Example 11.9, which was of order three) and may be impossible for orders greater than four except by strictly numerical techniques. What this means in practice is that stability studies are best done numerically for systems of even moderately high order. Computer routines exist for finding the eigenvalues of a square matrix. These routines give the $s_1$, $s_2$, ..., $s_M$ directly.

The picture is not yet so rosy for inverting ($s\mathbf{I}$ - $\mathbf{A}$) to find the transfer function matrix or the complete solution. Normal computer routines for matrix inversion are designed for numerical coefficients. The diagonal coefficients of ($s\mathbf{I}$ - $\mathbf{A}$) have the form ($s-a_{mm}$) so that the matrix inversion must be done algebraically rather than arithmetically. Computer programs capable of symbolic manipulation now exist, and it can be anticipated that they soon will be capable of inverting fairly large algebraic matrices and even capable of inverse Laplace transformation. In practice, however, most large sets of ODEs are solved numerically. The full equations usually contain nonlinearities which prevent analytical solution. The major use of the methods presented in this chapter is the analysis of control systems based on linearized versions of the equations. Often, the time-domain response is not required for this use.

## PROBLEMS

11.1 Find the Laplace domain and time domain transfer function matrices for the following system.

$$\frac{dx_1}{dt} = -x_1 - x_2 + u_1 + u_2$$

$$\frac{dx_2}{dt} = -x_2 + u_2$$

$$v_1 = x_1 + x_2 + u_1 + u_2$$

$$v_2 = x_2$$

11.2 Analyze the following set of equations with respect to stability.

$$\frac{dx_1}{dt} = -x_1 - x_3 + u_1$$

$$\frac{dx_2}{dt} = -x_2 - x_3 + u_1 + u_2$$

$$\frac{dx_3}{dt} = -x_1 - x_2 - x_3 + 2u_1$$

$$v = (x_1 + x_2 + x_3)/3$$

11.3 Suppose the output vector is a quadratic form in the state variables

$$\mathbf{v} = \mathbf{x}^T \mathbf{C}\, \mathbf{x}$$

How does this affect the general solution in Section 11.2?

11.4 Example 11.9 defined the matrix **B** as

$$\mathbf{B} = \begin{pmatrix} 0 & 0 \\ 0 & \alpha B \end{pmatrix}$$

How would the analysis change if

$$\mathbf{B} = \begin{pmatrix} 1 & 0 \\ 0 & \alpha\beta \end{pmatrix} ?$$

11.5 Does the system

$$\frac{dx_1}{dt} = -x_1 - x_2 - x_1 x_2$$

$$\frac{dx_2}{dt} = -x_2 + \alpha$$

have a stable steady state? Hint: Find all steady states; linearize the system about each steady state; perform a stability analysis.

# CHAPTER 12

# SIMULATION TECHNIQUES

Most of this book is devoted to analytical solutions of linear ODEs having constant coefficients. In the precomputer era, these were practically the only solutions possible. Now, numerical methods are available for solving large sets of nonlinear ODEs, and it is a rare problem that offers serious computational difficulties. However, merely coming up with a set of numbers provides little insight regarding the nature of the solution or of the real-world system being modeled. Most engineers find graphical displays far better than columns of numbers or even mathematical formulas.

One purpose of this chapter is to illustrate, in graphical form, the behavior of systems for which analytical solutions were developed in previous chapters. A second purpose is to introduce some general simulation techniques applicable to systems for which analytical solutions are infeasible. The same techniques work equally well for analytically solvable systems and typically produce numerical results faster than the analytical solution. The reason for this is that numerical solution techniques for ODEs tend to be fairly general. The same computational algorithm can be used for almost everything. Analytical solutions tend to be more case dependent. For example, the analytical solutions to

$$\frac{d^2x}{dt^2} + (\alpha+\beta)\frac{dx}{dt} + \alpha\beta x = \alpha\beta U(t) \tag{12.1}$$

depend on whether $\alpha$ and $\beta$ are real, imaginary, distinct, or repeated.

## 12.1 ODE SOLVERS

Essentially the first use of digital computers was to solve ODEs. Analog computers were developed in the 1960s with this sole function. They have since become obsolete, but at the time did have advantages with respect to graphical display of data and (arguably) ease of programming. Since the 1970s digital computers have re-emerged as the standard means for the numerical solution of sets of ODEs. The reasons for this are primarily the faster speeds and lower costs of digital computation. Improved computational algorithms and the advent of canned simulation packages have contributed to a lesser extent.

Hundreds of general purpose ODE solvers have been written. Some packages emphasize computational techniques and offer a variety of integration algorithms with such special features as automatic error control and step size adjustment. Typically, they are available as FORTRAN subroutines so that the user must supply the main program including I/O. Other packages emphasize a total systems approach where the user is spared programming details through a LOTUS-like environment. No one ODE solver has emerged as a standard, and it is unlikely that one will emerge in the near future. For this reason, we avoid reference to any canned packages. Instead, we will follow the practice used earlier in this book and give only the simplest possible algorithm for solving the example problems. The user can supply the I/O or can use one of the simulation packages.

### 12.1.1 Single First Order Equations

The most general first order ODE has the form

$$h\left(\frac{dx}{dt}, x, t\right) = 0 \tag{12.2}$$

where h is some function of the indicated variables. It is rarely, if ever, in problems of physical importance that Equation 12.2 cannot be solved explicitly for dx/dt. Thus, the most general problem typically reduces to

$$\frac{dx}{dt} = g(x,t) \tag{12.3}$$

where g is an arbitrary function of x and t. Chapter 1 showed how this equation can be solved by Euler's method, and Chapter 3 showed how it can be solved by the Runge-Kutta method. Neither method is apt to pose any difficulties.

Note that the system governed by Equation 12.3 need not be initially relaxed. The typical boundary condition for a first order system is

$$x = x_0 \text{ at } t = t_0$$

and the point $(x_0, t_0)$ is the usual starting point for the numerical integration. Occasionally, x' rather than x is specified at $t = t_0$, and it may be that $t_0 = \pm\infty$. Special solution techniques are then needed. These are seldom available in the canned packages but can be readily devised.

**EXAMPLE 12.1** The following equation arises in chemical reactor theory

$$\frac{da}{d\lambda} + \frac{4\lambda}{1 + 2\lambda}(1 - a) - ka^2 = 0$$

where $0 < a < 1$. The desired solution is the value for $a(\lambda)$ at $\lambda = 0$. The boundary condition is that $a' = 0$ at $\lambda = \infty$. Find $a_0 = a(0)$ using $k = 5$.

**Solution:** The boundary condition is unusual in two ways. It is specified in terms of a' rather than a, and it is specified at infinity, $\lambda_0 = \infty$. To solve the problem, values for $a_0$ will be assumed. Integration begins at the point $(a_0, 0)$ and continues until the long time behavior becomes apparent.

Euler's method is satisfactory for this problem. A basic program is

```
1.      INPUT A0
        A = A0
        T = 0
        N = 2048
        DT = 1/N
        FOR J = 1 to 2*N
        DA = -DT*(4*T/(1+2*T)*(1-A)-5*A*A)
        T = T + DT
        A = A + DA
        IF A>1 THEN PRINT "HIGH": GO TO 1
        IF A < 0 THEN PRINT "LOW": GO TO 1
        NEXT J
        PRINT "FINISH"
        GO TO 1
```

Running this program produces the following results:

```
Aϕ?        .50
HIGH
Aϕ?        .25
LOW
Aϕ?        .30
HIGH
Aϕ?        .27
LOW
Aϕ?        .28
LOW
Aϕ?        .29
HIGH
Aϕ         .285
HIGH
Aϕ?        .282
LOW
Aϕ?        .283
```

```
HIGH
Aϕ?          .2825
HIGH
Aϕ?
```

At this point, A0 is 0.282, accurate to three decimal places. The reader is invited to continue this game until "FINISH" is scored. This requires the input of A0 to high **precision**. The result will not be **accurate** to the same number of decimal positions due to the inaccuracy of Euler's method with DT = 1/2048.

More insight into the behavior of the solution can be found from a graphical representation. Plot $a(\lambda)$ versus $\lambda$ for the various values of $a_0$ considered above. This can be done using your favorite plotting routine, a canned simulation package, or the LINE command in BASIC. Whatever method is used, the results should look something like those in Figure 12.1.

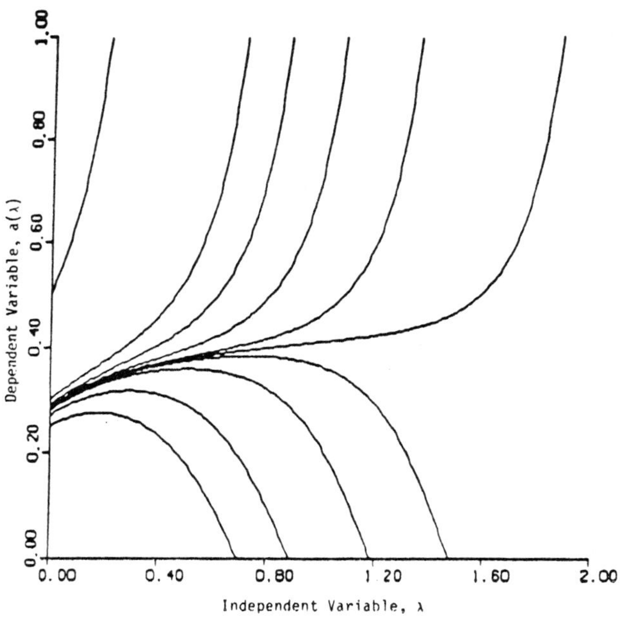


Figure 12.1 Example Solution of a Nonlinear, First Order ODE

It is apparent that only one value for $a_0$ satisfies the boundary condition. This value gives $a(\lambda)$ finite for all $\lambda$. Any other value gives $a(\lambda)$ diverging toward $\pm\infty$. An accurate trajectory for $a(\lambda)$ would be quite difficult to compute for large $\lambda$, but an accurate value for $a_0$ can be found easily.

## 12.1.2 Second Order Linear Systems

This section is intended to illustrate the general behavior of linear, second order ODEs with constant coefficients. Physical phenomena that can be modeled by such equations include the RCL circuits of Chapter 6 and the springs and dashpots of Chapter 8. Variable coefficients and nonlinear, second order equations will be considered in Section 12.1.3.

Equation 12.1 contains two parameters, $\alpha$ and $\beta$. Note that $-\alpha$ and $-\beta$ are the roots of the characteristic equation. These may be either real or imaginary. For computational purposes, it is much easier to deal with real parameters, yet we do not want to sacrifice the general behavior possible with Equation 12.1. We can have it both ways by replacing $\alpha$ and $\beta$ with two real-valued parameters

$$c_1 = -\alpha + -\beta \qquad \text{(sum of the roots)}$$

$$c_2 = \alpha\beta \qquad \text{(product of the roots)}$$

The governing equation becomes

$$\frac{dx^2}{dt^2} - c_1 \frac{dx}{dt} + c_2 x = c_2 U(t) \tag{12.4}$$

All possible values for $\alpha$ and $\beta$ are encompassed by real values of $c_1$ and $c_2$ in the range $-\infty$ to $+\infty$.

The relationship between $c_1$, $c_2$ and the roots can also be represented by

$$-\alpha = \frac{c_1 + \sqrt{c_1^2 - 4c_2}}{2}$$

$$-\beta = \frac{c_1 - \sqrt{c_1^2 - 4c_2}}{2}$$

Figure 12.2 shows how the roots, $-\alpha$ and $-\beta$, depend on $c_1$ and $c_2$. The roots are imaginary above the parabola $c_1^2 = 4c_2$ and real below the parabola. Stable solutions with $\alpha > 0$ and $\beta > 0$ are found only in the fourth quadrant. Within the fourth quadrant, the solution will be underdamped when $(c_1, c_2)$ lies above the parabola. It will be critically damped when $(c_1, c_2)$ lies on the parabola, and it will be underdamped when $(c_1, c_2)$ is below the parabola.

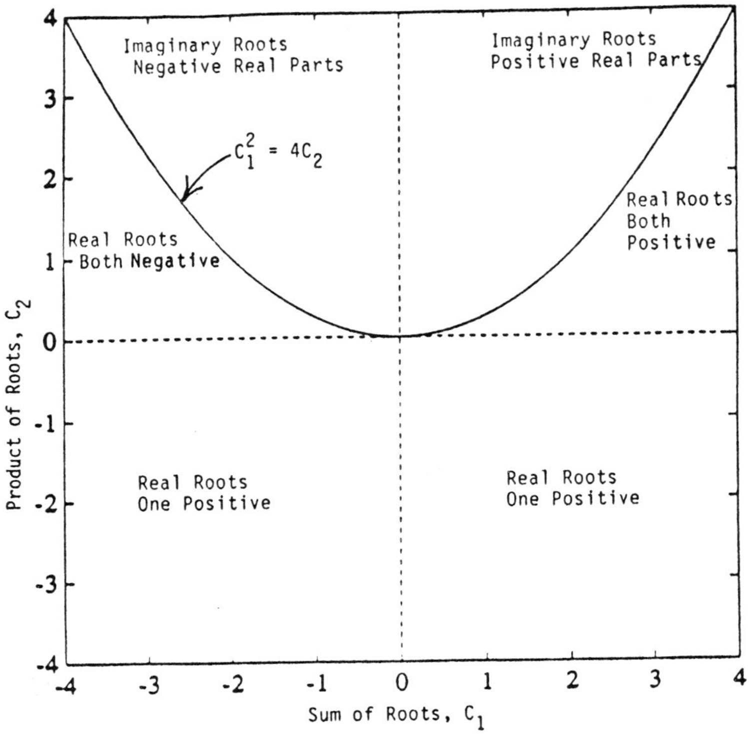


Figure 12.2 Roots of a Second Order Equation

**EXAMPLE 12.2** Simulate the behavior of

$$\frac{d^2x}{dt^2} - c_1\frac{dx}{dt} + c_2x = c_2U(t)$$

for a variety of $c_1$ and $c_2$. The system is initially relaxed.

**Solution:** We first express the governing ODE as a set of simultaneous, first order ODEs.

$$y = \frac{dx}{dt}$$

$$\frac{dy}{dt} - c_1y + c_2x = c_2U(t)$$

The initial conditions are $x = 0$ and $y = 0$ at $t = 0$. Euler's method is suitable for the numerical solution. A BASIC program segment for this solution is

```
INPUT "C1, C2"; C1,C2
N = 2048
DT = 1/N
T = 0
X = 0
Y = 0
FOR J = 1 to 8*N
DY = DT*(C2+C1*Y - C2*X)
DX = DT*X
T = T + DT
X = X + DX
Y = Y + DY
'PUT OUTPUT INSTRUCTIONS HERE
NEXT J
```

Figure 12.3 shows solutions when the point ($c_1$, $c_2$) is in the fourth quadrant. Only this quadrant gives stable solutions.

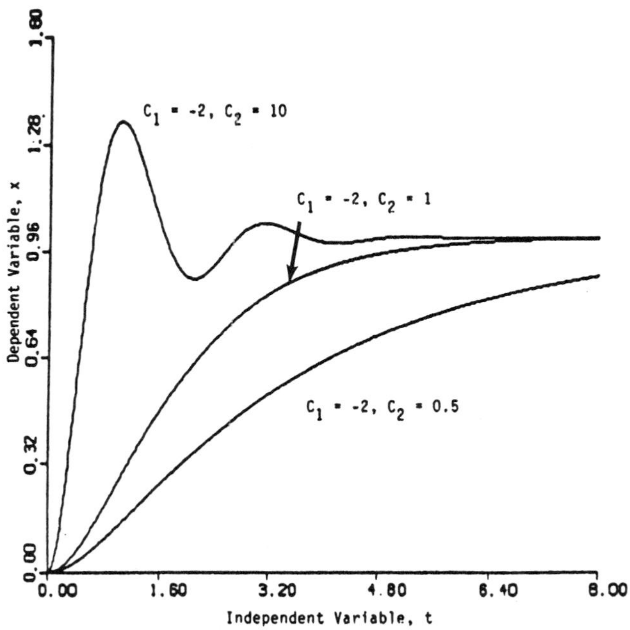


Figure 12.3 Solutions to a Second Order ODE With Both Roots Having Negative Real Parts

Figure 12.4 shows the various types of imaginary solutions. If $c_1 < 0$, the oscillation is damped and a steady state is approached. A sustained oscillation occurs when $c_1 = 0$, and a growing oscillation occurs when $c_1 > 0$.

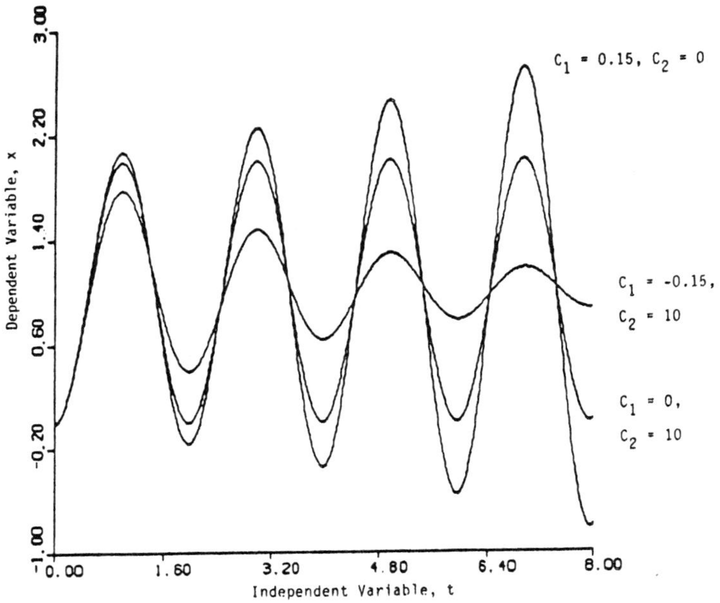


Figure 12.4 Solutions to a Second Order ODE With Imaginary Roots

---

**EXAMPLE 12.3** Simulate the response of the following feedback system when subjected to an input step change:

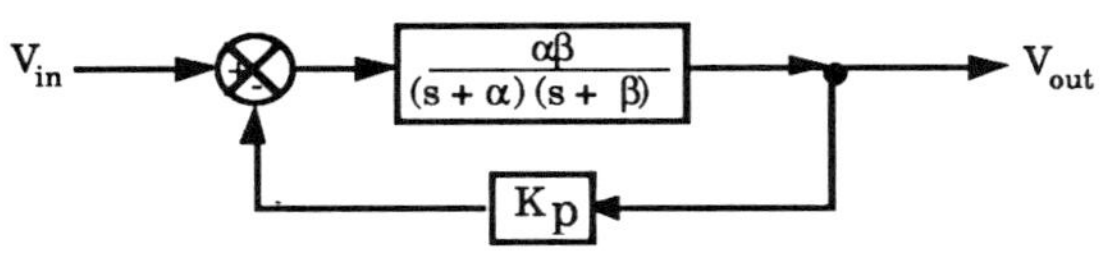


**Solution:** This is the problem studied in Example 9.6. For simulation purposes, we use the form of Equation 12.4 and add a term to the input which corresponds to negative, proportional feedback.

$$\frac{d^2v_{out}}{dt^2} - c_1 \frac{dv_{out}}{dt} + c_2 v_{out} = c_2[U(t) - K_p v_{out}] \qquad \text{(i)}$$

The system was relaxed prior to the step change, and the feedback signal tends to restore the initial state with $v_{out} = 0$. We proceed as in Example 12.2 and rewrite Equation i as a pair of simultaneous first order equations

$$\frac{dx}{dt} = y$$

$$\frac{dy}{dt} - c_1 y + c_2(1 + K_p)x = C_2 U(t)$$

where $x = v_{out}$. The BASIC program is essentially unchanged.

```
INPUT "KP,C1,C2"; KP,C1,C2
⋮
DY = DT*(C2+C1*Y - C2*(1+KP)*X)
⋮
```

Figure 12.5 shows the results for various values of $K_p$ of a simulation with $c_1 = -2$, and $c_2 = +0.75$. The open loop response is underdamped with $\alpha = 1.5$ and $\beta = 0.5$. $K_{crit} = 1/3$ for this system. See Example 9.6 for an analytical discussion of the simulation results.

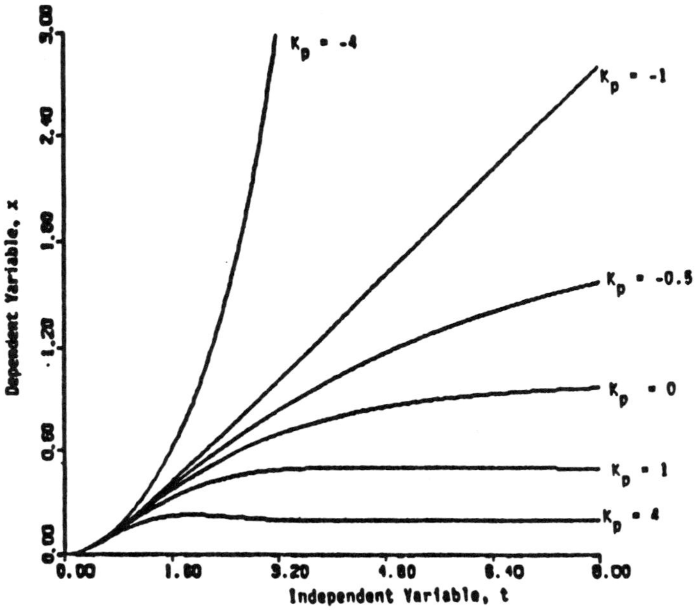


Figure 12.5 Response of a Second Order System Subject to Proportional Control: $C_1 = -2$, $C_2 = 0.75$.

**EXAMPLE 12.4** Add an integral term to the control signal in Example 12.3.

**Solution:** The governing equation is

$$\frac{d^2v_{out}}{dt^2} - c_1\frac{dv_{out}}{dt} + c_2v_{out} = c_2\,[U(t)-K_pv_{out} - K_I\int_0^t v_{out}dt] \qquad (i)$$

Equation i could be differentiated to give an equivalent third order system. For numerical simulation, it is easier to retain the second order structure and to evaluate the integral term numerically. The simulation equations are

$$\frac{dx}{dt} = y$$

$$\frac{dy}{dt} - c_1y + c_2(1 + K_p)x = c_2U(t) - c_2K_I\int_0^t xdt$$

The BASIC program segment is now

```
INPUT "KP, KI, C1, C2"; KP, KI, C1, C2
XINT = 0
:
XINT = XINT + X*DT
DY = DT*(C2+C1*Y-C2*(1+KP)*X - C2*KI*XINT)
```

Figure 12.6 shows results for a few values of $K_I$. The base case for these calculations was $c_1 = -2$, $c_2 = +.75$, and $K_p = 1$.

---

### 12.1.3 Simultaneous Nonlinear Equations

The previous section used two linear simultaneous, first order equations to solve a linear, second order equation. Adding additional first order equations to the set and even allowing them to be nonlinear does not necessarily make them difficult to solve numerically. It does enable much more complex dynamic behavior that can be difficult to analyze. Quite complex behavior is possible with a simple model, and one should wait a long time before reaching firm conclusions regarding stability. Even chaotic behavior with no readily discernible pattern can be generated by a small set of nonlinear ODEs.

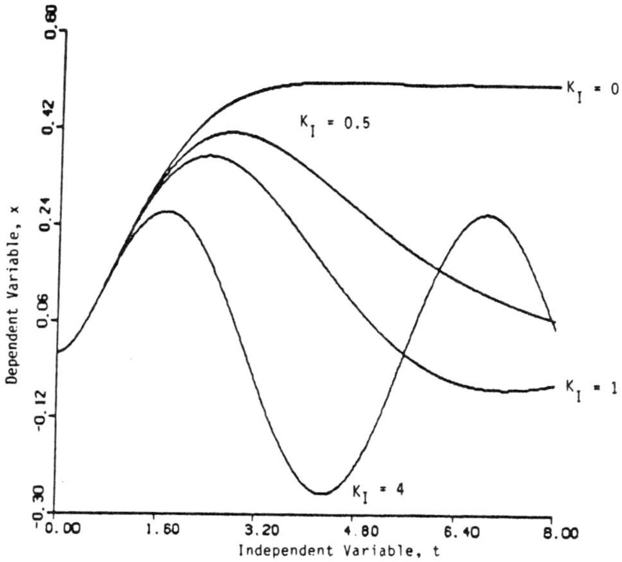


Figure 12.6 Response of a Second Order System Subject to Proportional and Integral Control: $C_1 = -2$, $C_2 = 0.75$, $K_P = 1$.

**EXAMPLE 12.5** Simulate the following set of ODEs.

$$\frac{dx}{dt} = .001xy + y - x^2 - 12 \cos t$$

$$\frac{dy}{dt} = .001xy + -0.1y - x^3 + 12 \cos t$$

The initial conditions are x = y =0 at t = 0.

**Solution:** The numerical solution is straightforward. The interesting thing about this example is how x and y vary with time. Figure 12.7 shows x vs. t at the start of the simulation with $0 < t < 8$. Figure 12.8 shows how the simulation ends at about t = 21. The simulation is obviously inaccurate near the end, but the general trend is clear. After a long time with complex but apparently bounded oscillations, x diverges in an oscillation of increasing frequency.

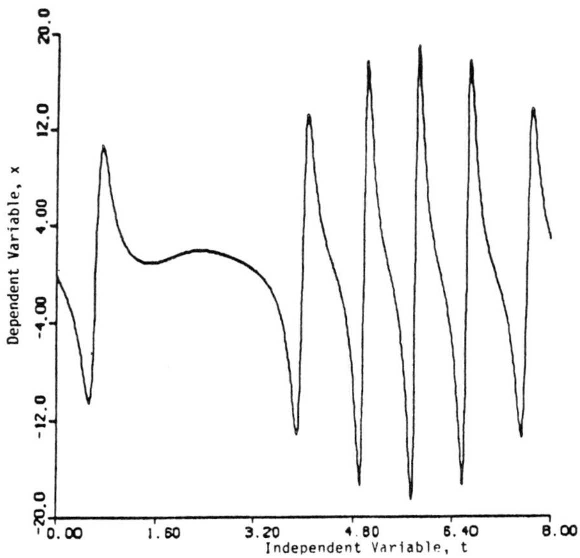


Figure 12.7 Initial Response of a Nonlinear, Second Order System.

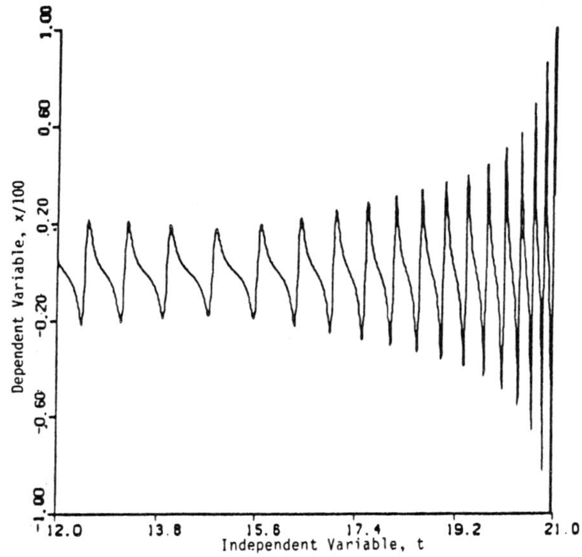


Figure 12.8 Long Time Response of a Nonlinear, Second Order System

## 12.2 MONTE CARLO TECHNIQUES

The models treated so far in this book have all involved differential equations. The model simulations have required numerical solutions of the governing ODEs. We turn now to a second type of model which avoids the direct use of differential equations and, indeed, of most formal mathematics. The model consists of a set of logical rules which we think the real system more or less follows and which we attempt to emulate using a digital computer. A prototype problem comes from **queueing theory**. Consider a turnpike exit with three toll booths. Cars arrive at a random rate. The service time needed to collect tolls, give directions, and so on is a random variable. Questions like what is the average waiting time in the queue and what is the maximum waiting time are quite difficult to answer by strictly analytical methods. They are fairly easily answered by computer simulation and the judicious use of statistics. Traffic management is not the subject of this book. Instead, we will demonstrate how Monte Carlo techniques can be used to solve a simple problem involving the residence time distribution in a stirred tank reactor.

**EXAMPLE 12.6** Develop a simulation model for the flow of molecules into and out of a perfectly mixed stirred tank. Use the model to determine the residence time washout function, w(t).

**Solution:** Chapter 10 used a strictly analytical approach to find that

$$w(t) = e^{-t/\bar{t}}$$

We now attempt to find this same result by a molecular level simulation.

Suppose there are M molecules in the stirred tank at any time. New molecules enter but old ones must leave to keep M constant. The rate at which molecules enter and leave the system is as fast as the computer will run. When M molecules have entered and left, one mean residence time, t, will have elapsed. The stirred tank is a perfect mixer by assumption. This means that each molecule has the same probability of leaving, regardless of when that molecule entered the system. The rules governing the simulation are simple:

1. Add a molecule to the system, recording its time of entry.

2. Remove a molecule from the system by random choice. Record its time of exit.

3. Calculate the residence time, t, for leaving molecules, this being the difference between exit time and entrance time.

4. Simulate enough events to develop good statistics on the distribution of the t. The washout function is the fraction of molecules leaving the system that had a residence time greater than t.

Following is a BASIC program that will perform the simulation

```
DIM MOLE (2048)
M = 2048

FOR I = 1 to M              'START MOLECULES WITH
MOLE(I) = 0                 'AN ENTRANCE TIME OF
NEXT I                          'ZERO

FOR I = 1 to 10*M           'RUN THE SIMULATION
K = RND                     'FOR 10 RESIDENCE
MOLE (K) = I                'TIMES TO APPROACH
NEXT I                      'STEADY STATE

FOR I = 10*M+1 to 11*M      'NOW SAMPLE FOR
K = RND                     'AN 11TH RESIDENCE
T = (I-MOLE(K))/M           'TIME
MOLE (K) = I                'COLLECT STATISTICS ON THE T
NEXT I

                            'OUTPUT RESULTS
```

Results for this simulation are

| $t/\bar{t}$ | w(t) by Simulation | w(t) by Theory |
|---|---|---|
| .1 | .893 | .905 |
| .2 | .801 | .819 |
| .3 | .727 | .741 |
| .4 | .667 | .670 |
| .5 | .610 | .607 |
| .6 | .559 | .549 |
| .7 | .511 | .497 |
| .8 | .459 | .449 |
| .9 | .414 | .407 |
| 1.0 | .377 | .368 |
| 2.0 | .131 | .135 |
| 3.0 | .058 | .050 |
| 4.0 | .019 | .018 |
| 5.0 | .006 | .007 |
| 6.0 | .003 | .002 |

The foregoing example simulated events on a molecular scale. This is the classical Monte Carlo approach. Although the example program is simple and runs very fast, more typical molecular simulations require enormous amounts of computing time. Normally, they are used only for problems which cannot be formulated in terms of conventional mathematics.

The Monte Carlo simulation is frequently used to find total processing times and queue lengths in multistep, multimachine manufacturing operations. The general approach is called **discrete event simulation**. It simulates a composite system using rules which govern the individual parts. It is analogous to the determination of composite transfer functions by the network analysis technique of Chapter 9. We treat a simple example readily solved by other means.

**EXAMPLE 12.7** Determine the composite washout function, $w_c(t)$, for two equal volume stirred tanks in series.

**Solution:** The individual washout functions are

$$w_1 = e^{t_1/\bar{t}}$$

$$w_2 = e^{t_2/\bar{t}}$$

where $\bar{t}$ is the mean residence time in a single tank.

This problem can be solved analytically using the methods of Chapters 9 and 10. The result is

$$w_c(t) = e^{-t/\bar{t}}\,[1 + t/\bar{t}]$$

$$= e^{-\tau}\,[1 + \tau]$$

where $\tau = t/\bar{t}$. It could also be solved using molecular simulation as in Example 12.6. There would be two tanks. A molecule leaving the first tank would enter the second and would continue to accumulate residence time. Statistics on t would be gathered at the exit of the second tank. A third approach, to be illustrated below, has the spirit of a discrete event simulation. We sample $t_1$ using the washout function for the first tank. We independently sample $t_2$ using the washout function for the second tank. Then $t_c = t_1 + t_2$ constitutes a residence time sample for the composite system.

Sampling a population is quite easy when w(t) is known. A random number in the range 0 to 1 is used for w and the corresponding value for t is calculated. For the case at hand, t can be found analytically as a function of w:

$$\tau_1 = \frac{t_1}{\bar{t}} = -\ln w_1$$

$$\tau_2 = \frac{t_2}{\bar{t}} = -\ln w_2$$

A BASIC program segment is

```
LIM = 2048
FOR I = 1 to LIM
W1 = RND
TAU1 = -LOG(W1)
W2 = RND
TAU2 = -LOG(W2)
TAU = TAU1 + TAU2
'COLLECT STATISTICS ON THE TAU
NEXT I
```

Some results are

| $\tau$ | $w(\tau)$ by Simulation | w(t) by Theory |
|---|---|---|
| .1 | .989 | .995 |
| .2 | .976 | .982 |
| .5 | .897 | .910 |
| 1.0 | .732 | .736 |
| 2.0 | .419 | .406 |
| 3.0 | .203 | .199 |
| 4.0 | .090 | .092 |
| 5.0 | .042 | .040 |
| 6.0 | .014 | .017 |

---

## PROBLEMS

12.1. Given

$$\frac{dx}{dt} = -x\,(1 - 0.1/t)$$

subject to x = 1 at t = 0, find the maximum value of x.

12.2. Given

$$\frac{dx}{dt} = \exp\left(-t\frac{dx}{dt}\right)$$

subject to $x = 0$ at $t = 0$, find $x$ at $t = 8$. This problem may lack physical significance but it is well posed mathematically.

12.3. Duplicate Figure 12.5 for the case of $c_1 = -2$ and $c_2 = 1$.

12.4. Simulate the following set of ODEs.

$$\frac{dx}{dt} = 2.1x - xy$$

$$\frac{dy}{dt} = -y + xy$$

subject to $x = y = 1$ at $t = 0$. Specifically, find the minimum and maximum values for x and y.

12.5. Use Monte Carlo techniques to find w(t) for the following flow system.

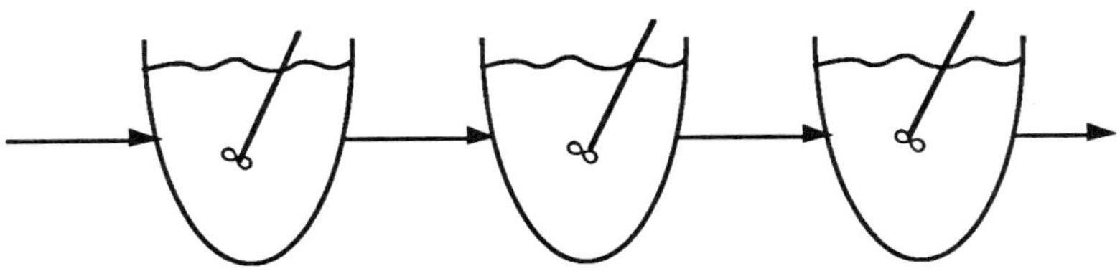

Assume the stirred tanks have equal volumes.

12.6. Use Monte Carlo techniques to find w(t) for the following flow system.

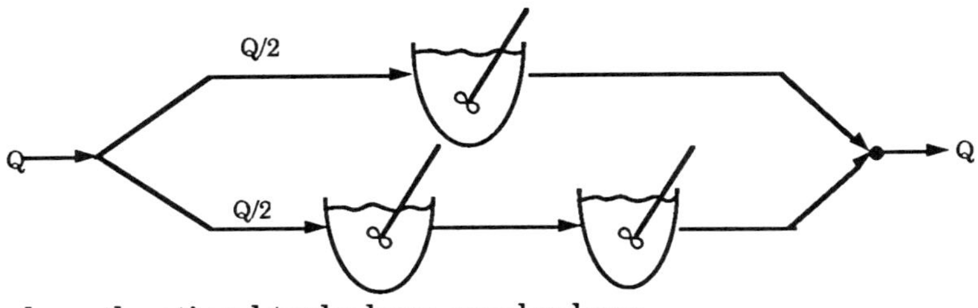


where the stirred tanks have equal volume.

12.7. Add derivative action to the controller in Example 12.4. Simulate the results for various $K_D$ using $C_1 = -2$, $C_2 = -.75$, $K_p = 1$ and $K_I = 1$.

# INDEX